Joel Dorman Steele

Hygienic physiology

with special reference to the use of alcoholic drinks and narcotics

Joel Dorman Steele

Hygienic physiology
with special reference to the use of alcoholic drinks and narcotics

ISBN/EAN: 9783744737432

Printed in Europe, USA, Canada, Australia, Japan

Cover: Foto ©berggeist007 / pixelio.de

More available books at **www.hansebooks.com**

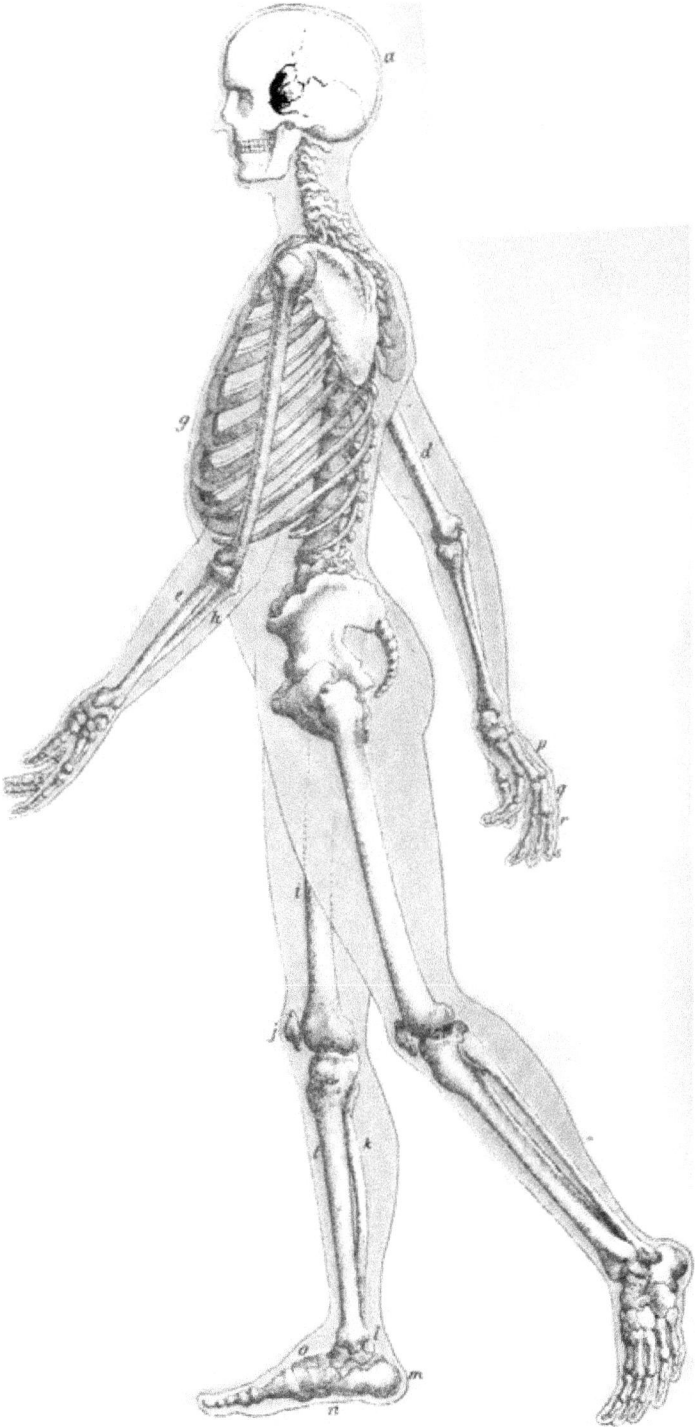

Hygienic Physiology,

WITH SPECIAL REFERENCE TO THE USE OF

ALCOHOLIC DRINKS AND NARCOTICS.

ADAPTED FROM THE

"FOURTEEN WEEKS IN HUMAN PHYSIOLOGY,"

BY

JOEL DORMAN STEELE, Ph.D.

Edited and Endorsed for the use of Schools (in accordance with the recent Legislation upon this subject) by the Department of Scientific Temperance Instruction of the W. C. T. U. of the United States, under the direction of Mrs. Mary H. Hunt, Supt.

A. S. BARNES & COMPANY,

NEW YORK AND CHICAGO

THE FOURTEEN-WEEKS SERIES

IN

NATURAL SCIENCE,

BY

J. DORMAN STEELE, Ph.D., F.G.S.

New Physics.

New Chemistry.

New Descriptive Astronomy.

Popular Geology.

Human Physiology.

Zoology.

Botany.

A Key, containing Answers to the Questions and Problems in Steele's
14-Weeks Series.

AN HISTORICAL SERIES,

ON THE PLAN OF STEELE'S 14-WEEKS IN THE SCIENCES.

A Brief History of the United States.

A Brief History of France.

A Brief History of Ancient Peoples.

A Brief History of Mediæval and Modern Peoples.

A Brief General History.

A Brief History of Greece.

A Brief History of England.   (In preparation.)

A Popular History of the United States.

PREFACE.

THE term Physiology, or the science of the functions of the body, has come to include Anatomy, or the science of its structure, and Hygiene, or the laws of health; the one being essential to the proper understanding of physiology, and the other being its practical application to life. The three are intimately blended, and in treating of the different subjects the author has drawn no line of distinction where nature has made none. This work is not prepared for the use of medical students, but for the instruction of youth in the principles which underlie the preservation of health and the formation of correct physical habits. All else is made subservient to this practical knowledge. A simple scientific dress is used which, while conducing to clearness, also gratifies that general desire of children to know something of the nomenclature of any study they pursue.

To the description of each organ is appended an account of its most common diseases, accidents, etc., and, when practicable, their mode of treatment. A pupil may thus learn, for example, the cause and cure of a cold, the management of a wound, or the nature of an inflammation.

The Practical Questions, which have been a prominent feature of the series, will be found, it is hoped, equally useful in this work. Directions for preparing simple microscopic objects, and illustrations of the different organs, are given under each subject.

In the Appendix will be found Questions for class use, Hints about the sick-room, Suggestions as to what to do "till the doctor comes," Antidotes for poisons, and a full Index.

Believing in a Divine Architect of the human form, the author cannot refrain from occasionally pointing out His inimitable workmanship, and impressing the lesson of a Great Final Cause.

The author has gleaned from every field, at home and abroad, to secure that which would interest and profit his pupils. In general, Flint's great work on the Physiology of Man, an undisputed authority on both sides of the Atlantic, has been adopted as the standard in digestion, respiration, circulation, and the nervous system. Leidy's Human Anatomy, and Sappey's Traité d'Anatomie have been followed on all anatomical questions, and have furnished many beautiful drawings. Huxley's Physiology has af-

forded exceedingly valuable aid. Foster's Text-
Book of Physiology, Hinton's delightful work on
Health and its Conditions, Black's valuable Ten
Laws of Health, Williams's practical essay on Our
Eyes and How to Use them, Le Pileur's charming
treatise on The Wonders of the Human Body, and
that quaint volume, Odd Hours of a Physician, have
aided the author with facts and fancies. The writ-
ings of Draper, Dalton, Carpenter, Valentine, Mapother, Watson, Lankester, Letheby, Hall, Hamil-
ton, Bell, Wilson, Bower, Cutter, Hutchison, Wood,
Bigelow, Stillé, Holmes, Beigel and others have been
freely consulted.

READING REFERENCES.

FOSTER's Text-book of Physiology.—Leidy's Human Anatomy.—Draper's Human Physiology.—Dalton's Physiology and Hygiene.—Cutter's Physiology.—Johnston & Church's Chemistry of Common Life.—Letheby's Food.—Tyndall on Light, and on Sound.—Flint's Physiology of Man.—Rosenthal's Physiology of the Muscles and Nerves.—Bernstein's Five Senses of Man.—Huxley & Youmans's Physiology & Hygiene.—Sappey's Traité d'Anatomie.—Luys's Brain and its Functions.—Smith's Foods.—Bain's Mind and Body.—Pettigrew's Animal Locomotion. — Carpenter's Mental Physiology.—Wilder and Gage's Anatomy.

Hargreave's Alcohol and Science.—Richardson's Ten Lectures on Alcohol, and Diseases of Modern Life. — Brown's Alcohol.—Davis's Intemperance and Crime.— Pitman's Alcohol and the State.—Anti-Tobacco.—Howie's Stimulants and Narcotics. —Hunt's Alcohol as Food or Medicine.—Schützenberger's Fermentation.

SUGGESTIONS TO TEACHERS.

SEEING is believing, — more than that, it is often knowing and remembering. The mere reading of a statement is of little value compared with the observation of a fact. Every opportunity should therefore be taken of exhibiting to the pupil the phenomena described, and thus making them real. A microscope is so essential to the understanding of many subjects, that it is indispensable to the proper teaching of Physiology. A suitable instrument and carefully prepared specimens showing the structure of the bones, the skin, and the blood of various animals, the pigment cells of the eye, etc., may be obtained at a small cost from the Publishers of this book.

On naming the subject of a paragraph, the pupil should be prepared to tell all he knows about it. No failure should discourage the teacher in establishing this mode of study and recitation. A little practice will produce the most satisfactory results.

The unexpected question and the apt reply develop a certain sharpness and readiness which are worthy of cultivation. The questions for review, or any others that the wit of the teacher may suggest, can be effectively used to break the monotony of a topical recitation, thereby securing the benefits of both systems.

The pupil should expect to be questioned each day upon any subject passed over during the term, and thus the entire knowledge gained will be within his grasp for instant use. While some are reciting to the teacher, let others write on slates or on the blackboard. At the close of the recitation let all criticise the ideas, the spelling, the use of capitals, the pronunciation, the grammar, and the mode of expression. Greater accuracy and much collateral drill may thus be secured at little expense of valuable school-time.

The Introduction is designed merely to furnish suggestive material for the first lesson, preparatory to beginning the study. Other topics may be found in the questions given in the Appendix. In this same connection read also the Conclusion.

TABLE OF CONTENTS.

V I I I .

I X .

X .

INTRODUCTION.

PHYSIOLOGICAL STUDY in youth is of inestimable value. Precious lives are frequently lost through ignorance. Thousands squander in early years the strength which should have been kept for the work of real life. Habits are often formed in youth which entail weakness and poverty upon manhood, and are a cause of life-long regret. The use of a strained limb may permanently damage it. Some silly feat of strength may produce an irreparable injury. A thoughtless hour of reading by twilight may impair the sight for life. A terrible accident may happen, and a dear friend perish before our eyes, while we stand by powerless to render the assistance we could so easily give did we "only know what to do." The thousand little hints which may save or lengthen life, may repel or abate disease, and the simple laws which regulate our bodily vigor, should be so familiar that we may be quick to apply them in an emergency. The preservation of health is easier than the cure of disease. Childhood cannot afford to wait for the lesson of experience which is learned only when the penalty of violated law has been already incurred, and health irrevocably lost.

Nature's Laws Inviolable. — In infancy, we learn how terribly Nature punishes a violation of certain laws, and how promptly she applies the penalty. We soon find out the peril of fire, falls, edged-tools, and the like. We fail, however, to notice the equally sharp and certain punishments which bad habits entail. We are quick to feel the need of food, but not so ready to perceive the danger of an excess. A lack of air drives us at once to secure a supply; but foul air is as fatal, yet gives us no warning.

Nature provides a little training for us at the outset of life, but leaves the most for us to learn by bitter experience. So in youth we throw away our strength as if it were a burden of which we desired to be rid. We eat anything, and at any time; do anything we please, and sit up any number of nights with little or no sleep. Because we feel only a momentary discomfort from these physical sins, we fondly imagine when that is gone we are all right again. Our drafts upon our constitution are promptly paid, and we expect this will always be the case; but some day they will come back to us protested; Nature will refuse to meet our demands, and we shall find ourselves physical bankrupts.

We are furnished in the beginning with a certain vital force upon which we may draw. We can be spendthrifts and waste it in youth, or be wise and so husband it to manhood. Our shortcomings are all charged against this stock. Nature's memory never fails : she keeps her account with perfect exactness. Every physical sin subtracts from the sum and strength of our years. We may cure a disease, but it never leaves us as it found us. We may heal a wound, but the scar still shows. We reap as we sow, and we may either gather in the thorns, one by one, to torment and destroy, or rejoice in the happy harvest of a hale old age.

I.

THE SKELETON.

" Not in the World of Light alone,
Where God has built His blazing throne,
Nor yet alone on earth below,
With belted seas that come and go,
And endless isles of sunlit green
Is all thy Maker's glory seen—
Look in upon thy wondrous frame,
Eternal wisdom still the same !"

HOLMES.

ANALYSIS OF THE SKELETON.

NOTE.—The following Table of 206 bones is exclusive of the 8 sesamoid bones which occur in pairs at the roots of the thumb and great toe, making 214 as given by Leidy, and Draper. Gray omits the bones of the ear, and names 200 as the total number.

THE SKELETON.

I. THE HEAD. (28 bones.)

1. CRANIUM (8 bones).
- Frontal bone (forehead).
- Two Parietal bones.
- Two Temporal (temple) bones.
- Sphenoid bone.
- Ethmoid (sieve-like bone at root of nose).
- Occipital bone (back and base of skull).

2. FACE (14 bones.)
- Two Superior Maxillary (upper jaw) bones.
- Inferior Maxillary (lower jaw) bone.
- Two Malar (cheek) bones.
- Two Lachrymal bones.
- Two Turbinated (scroll-like) bones, each side of nose.
- Two Nasal bones (bridge of nose).
- Vomer (the bone between the nostrils).
- Two Palate bones.

3. EARS (6 bones.)
- Hammer.
- Anvil.
- Stirrup.

II. THE TRUNK. (54 bones.)

1. SPINAL COLUMN
- Cervical Vertebræ (seven vertebræ of the neck).
- Dorsal Vertebræ (twelve vertebræ of the back).
- Lumbar Vertebræ (five vertebræ of the loins).

2. RIBS
- True Ribs.
- False Ribs.

3. STERNUM (breast-bone).

4. Os HYOIDES (bone at the root of tongue).

5. PELVIS
- Two Innominata.
- Sacrum.
- Coccyx.

III. THE LIMBS. (124 bones.)

1. UPPER LIMBS (64 bones.)
- *Shoulder* — Clavicle. Scapula.
- *Arm* — Humerus. Ulna and Radius.
- *Hand* — Eight Wrist or Carpal bones. Five Metacarpal bones. Phalanges (14 bones).

2. LOWER LIMBS (60 bones.)
- *Leg* — Femur. Patella. Tibia and Fibula.
- *Foot* — Seven Tarsal bones. Five Metatarsal bones. Phalanges (14 bones).

THE SKELETON.

1. FORM, STRUCTURE, ETC. OF THE BONES.
1. Uses.
2. Composition.
3. Structure.
4. Growth.
5. Repair.
6. The Joints.

2. CLASSIFICATION OF THE BONES.
1. The Head.
2. The Trunk.
3. The Limbs.

THE SKELETON.

THE **Skeleton,** or framework, of the "House we live in," is composed of about 200 bones.*

Uses and Forms of the Bones.—They have three principal uses : 1. To protect the delicate organs;† 2. To serve as levers on which the muscles may act to produce motion ; and 3. To preserve the shape of the body.

Bones differ in form according to the uses they subserve. For convenience in walking, some are long ; for strength and compactness, some are short and thick ; for covering a cavity, some are flat ; and for special purposes, some are irregular. The general form is such as to combine strength and lightness. For example, all the long bones of the limbs are round and hollow, thus giving with the same

---

* The precise number varies in different periods of life. Several which are separated in youth become united in old age. Thus five of the " false vertebræ" at the base of the spine early join in one great bone—the sacrum ; while four tiny ones below it often run into a bony mass—the coccyx (Fig. 6); in the child, the sternum is composed of eight pieces, while in the adult it consists of only three. While, however, the number of the bones is uncertain, their relative length is so exact that the length of the entire skeleton, and thence the height of the man, can be obtained by measuring a single one of the principal bones. Fossil bones and those found at Pompeii have the same proportion as our own.

† An organ is a portion of the body designed for a particular use, called its *function.* Thus the heart circulates the blood ; the liver produces the bile.

weight a greater strength,* and also a larger surface for the attachment of the muscles.

The Composition of the Bones at maturity is about one part animal to two parts mineral matter. The proportion varies with the age. In youth it is nearly half and half, while in old age the mineral is greatly in excess. By soaking a bone in weak muriatic acid, and thus dissolving the mineral matter, its shape will not change, but its stiffness will disappear, leaving a tough, gristly substance† (cartilage), which can be bent like rubber.

If the bone be burned in the fire, thus consuming the animal matter, the shape will still be the same, but it will have lost its tenacity, and the beautiful, pure-white residue‡ may be crumbled into powder with the fingers.

We thus see that a bone receives hardness and rigidity from its mineral, and tenacity and elasticity from its animal matter.

* Cut a sheet of foolscap in two pieces. Roll one-half into a compact cylinder, and fold the other into a close, flat strip; support the ends of each, and hang weights in the middle until they bend. The superior strength of the roll will astonish one unfamiliar with this mechanical principle. In a rod, the particles break in succession, first those on the outside, and later those in the center. In a tube, the particles are all arranged where they resist the first strain. Iron pillars are therefore cast hollow. Stalks of grass and grain are so light as to bend before a breath of wind, yet are stiff enough to sustain their load of seed.

† Mix a wineglass of muriatic acid with a pint of water, and place in it a sheep's rib. In a day or two. it will be so soft that it can be tied into a knot. In the same way, an egg may be made so pliable that it can be crowded into a narrow-necked bottle, within which it will expand, and become an object of great curiosity to the uninitiated. By boiling bones at a high temperature, the animal matter separates in the form of gelatin. Dogs and cats extract the animal matter from the bones they eat. Fossil bones deposited in the ground during the Geologic period, were found by Cuvier to contain considerable animal matter. Gelatin was actually extracted from the Cambridge mastodon, and made into glue. A tolerably nutritious food might thus be manufactured from bones older than man himself.

‡ From bones thus calcined, the phosphorus of the chemist is made. See *Chemistry*, page 120. If the animal matter be not consumed, but only charred, the bone will be black and brittle. In this way, the "bone-black" of commerce is manufactured.

Fig. 2.

The entire bone is at first composed of cartilage, which gradually *ossifies* or turns to bone.* Certain portions near the joints are long delayed in this process, and by their elasticity assist in breaking the shock of a fall.† Hence the bones of children are tough, are not readily fractured, and when broken easily heal again;‡ while those of elderly people are liable to fracture, and do not quickly unite.

The Structure of the Bones.—When a bone is sawed lengthwise, it is found to be a compact shell filled with a spongy substance. This filling increases in quantity, and becomes more porous

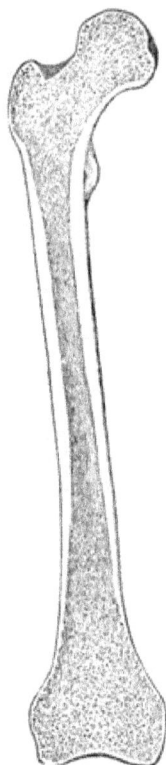

The thigh-bone, or femur sawed lengthwise.

* "The ossification of the bones on the sides and upper part of the skull, for example, begins by a rounded spot in the middle of each one. From this spot the ossification extends outward in every direction, thus gradually approaching the edges of the bone. When two adjacent bones meet, there will be a line where their edges are in contact with each other, but have not yet united ; but when more than two bones meet in this way, there will be an empty space between them at their point of junction. Thus, if you lay down three coins upon the table with their edges touching one another, there will be a three-sided space in the middle between them ; if you lay down four coins in the same manner, the space between them will be four-sided. Now at the back part of the head there is a spot where three bones come together in this way, leaving a small, three-sided opening between them : this is called the "posterior fontanelle." On the top of the head four bones come together, leaving between them a large, four-sided opening : this is called the "anterior fontanelle." These openings are termed the *fontanelles*, because we can feel the pulsations of the brain through them, like the bubbling of water in a fountain. The fontanelles gradually diminish in size, owing to the growth of the bony parts around them, and are completely closed at the age of four years after birth."—*Dalton's Physiology*, p. 361.

† Frogs and toads, which move by jumping, and consequently receive so many jars, retain these unossified portions (epiphyses) nearly through life ; while alligators and turtles, whose position is sprawling, and whose motions are measured, do not have them at all.—*Leidy*.

‡ This is only one of the many illustrations of the Infinite care that watches over helpless infancy, until knowledge and ability are acquired to meet the perils of life.

at the ends of the bone, thus giving greater size to form a strong joint, while the solid portion increases near the middle, where strength alone is needed. Each fiber of this bulky material diminishes the shock of a sudden blow, and also acts as a beam to

Fig. 3.

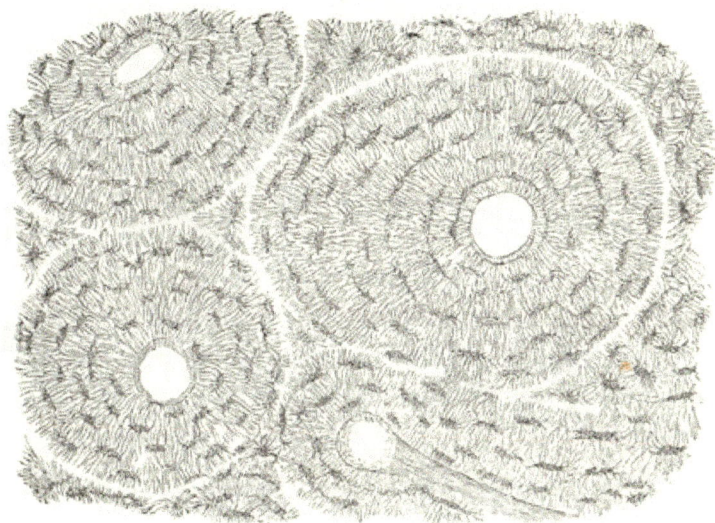

A thin slice of bone, highly magnified, showing the lacunæ, the tiny tubes (canaliculi) radiating from them, and four Haversian canals, three seen crosswise and one lengthwise.

brace the exterior wall. The recumbent position of the alligator protects him from falls, and therefore his bones contain very little spongy substance.

In the body, bones are not the dry, dead, blanched things they commonly seem to be, but are moist, living, pinkish structures, covered with a tough membrane called the per-i-os'-te-um.* (*peri*, around,

* The relations of the periosteum to the bone are very interesting. Instances are on record where the bone has been removed, leaving the periosteum, from which the entire bone was afterward renewed.

and *osteon*, a bone), while the hollow is filled with marrow, rich in fat, and full of blood-vessels. If we examine a thin slice with the microscope, we shall see black spots with lines running in all directions, and looking very like minute insects. These are really little cavities called *la-cú-næ*,* from which radiate tiny tubes. The lacunæ are arranged in circles around larger tubes, termed from their discoverer, *Haversian canals*, which serve as passages for the blood-vessels that nourish the bone.

Growth of the Bones.—By means of this system of canals, the blood circulates as freely through the bones as through any part of the body. The whole structure is constantly but slowly changing,† old material being taken out and new put in. A curious illustration is seen in the fact that if madder be mixed with the food of pigs, it will tinge their bones red.

Repair of the Bones.—When a bone is broken, the blood at once oozes out of the fractured ends. This soon gives place to a watery fluid, which in a fortnight thickens to a gristly substance strong enough to hold them in place. Bone-matter is then slowly deposited, which in five or six weeks will unite the broken parts. Nature, at first, apparently endeavors to remedy the weakness of the material by excess in the quantity, and so the new portion is larger than the old. But the extra matter will be gradually

---

* When the bone is dry, the lacunæ are filled with air, which refracts the light, so that none of it reaches the eye, and hence the cavities appear black.

† Bone is sometimes produced with surprising rapidity. The great Irish Elk is calculated by Prof. Owen to have cast off and renewed annually in its antlers eighty pounds of bone.

absorbed, sometimes so perfectly as to leave no trace
of the injury.

A broken limb should always be held in place
by splints to enable this process to go on uninter-
ruptedly, and also lest a sudden jar might rupture
the partially-mended break. For a long time, the
new portion consists largely of animal matter, and
so is tender and pliable. The utmost care is therefore necessary to prevent a malformation.

The Joints are packed with a soft, smooth cartilage, or gristle, which fits so perfectly as often to be
air-tight. Upon convex surfaces, it is thickest at the
middle, and upon concave surfaces, it is thickest at
the edge, or where the wear is greatest. In addition, the ends of the bones are covered with a thin
membrane, the *synovial* (*sun*, with ; *ovum*, an egg),
which secretes a viscid fluid, not unlike the white of
an egg. This lubricates the joints, and prevents the
noise and wear of friction. The body is the only
machine that oils itself.

The bones which form the joint are tied with stout
ligaments (*ligo*, I bind), or bands, of a smooth, silvery
white tissue,* so strong that the bones are sometimes
broken without injuring the fastenings.

* The general term *tissue* is applied to the various textures of which the organs
are composed. For example, the osseous tissue forms the bones ; the fibrous tissue,
the skin, tendons, and ligaments.

II. CLASSIFICATION OF THE BONES.

For convenience, the bones of the skeleton are considered in three divisions : the *head*, the *trunk* and the *limbs*.

1. THE HEAD.

Fig. 4.

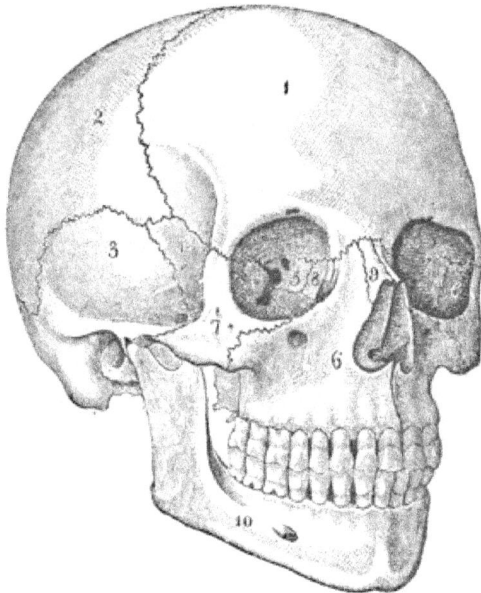

The Skull.—1, frontal bone ; 2, parietal bone ; 3, temporal bone ; 4, the sphenoid bone ; 5. ethmoid bone ; 6, superior maxillary (upper jaw) bone; 7, malar bone ; 8, lachrymal bone ; 9, nasal bone ; 10, inferior maxillary (lower jaw) bone.

The Bones of the Skull and the Face form a cavity for the protection of the brain and the four organs of sense, viz. : sight, smell, taste, and hearing. All of these bones are immovable except the lower jaw, which is hinged* at the back so as to allow for the opening and shutting of the mouth.

The Skull is composed, in general, of two compact plates, with a spongy layer between. These are in several pieces, the outer ones being joined by notched edges (sutures, sūt'yurs) in the way carpenters term dove-tailing. (See Fig. 4.)

The peculiar structure and form of the skull afford a perfect shelter for the brain—an organ so delicate that, if unprotected, an ordinary blow would destroy it. Its oval or egg shape adapts it to resist pressure. The smaller and stronger end is in front, where the danger is greatest. Projections before and behind shield the less protected parts. The hard plates are not easy to penetrate.† The spongy packing deadens every blow.‡ The separate pieces with their curious joinings disperse any jar which one may receive, and also prevent fractures from spreading.

The frequent openings in this strong bone-box afford safe avenues for the passage of numerous nerves and vessels which communicate between the brain and the rest of the body.

---

* A ring of cartilage is inserted in its joints, something after the manner of a washer in machinery. This follows the movements of the jaw, and admits of freer motion, while it guards against dislocation.

† Instances have been known where bullets striking against the skull have glanced off, been flattened, or even split into halves. In the Peninsular Campaign, the author saw a man who had been struck in the forehead by a bullet which, instead of penetrating the brain, had followed the skull around to the back of the head, and there passed out.

‡ An experiment resembling the familiar one of the balls in
Fig. 5. Natural Philosophy (*Steele's Physics*, Fig. 7, p. 30), beautifully illustrates this point. Several balls of ivory are suspended by cords, as in Fig. 5. If A be raised and then let fall, it will transmit the force to B, and that to C, and so on until F is reached, which will fly off with the impulse. If now a ball of spongy bone be substituted for an ivory one anywhere in the line, the force will be checked, and the last ball will not stir.

2. THE TRUNK.

The Trunk has two important cavities. The upper part, or *chest*, contains the heart and the lungs, and the lower part, or *abdomen*, holds the stomach, liver, kidneys, and other organs (Fig. 31). The principal bones are those of the *spine*, the *ribs*, and the *hips*.

The Spine consists of twenty-four bones, between which are placed pads of cartilage.* A canal is hollowed out of the column for the safe passage of the spinal cord. (See Fig. 50.) Projections (processes) at the back and on either side are abundant for the attachment of the muscles. The packing acts as a cushion to prevent any jar from reaching the brain when we jump or run, while the double curve of the spine also tends to disperse the force of a fall. Thus on every side the utmost caution is taken to guard that precious gem in its casket.

The Perfection of the Spine surpasses all human contrivances. Its various uses seem a bundle of con-

Fig. 6.

The Spine; the seven vertebræ of the neck, cervical; the twelve of the back, dorsal; the five of the loins, lumbar; a, the sacrum, and b, the coccyx. comprising the nine "false vertebræ" (p. 5).

---

* These pads vary in thickness from one-fourth to one-half an inch. They become condensed by the weight they bear during the day, so that we are somewhat shorter at evening than in the morning. Their elasticity causes them to resume their usual size during the night, or when we lie down for a time.

tradictions. A chain of twenty-four bones is made
so stiff that it will bear a heavy burden, and so flexible that it will bend like rubber ; yet, all the while,
it transmits no shock, and even hides a delicate nerve
within that would thrill with the slightest touch.
Resting upon it, the brain is borne without a tremor ;
and, clinging to it, the vital organs are carried without fear of harm.

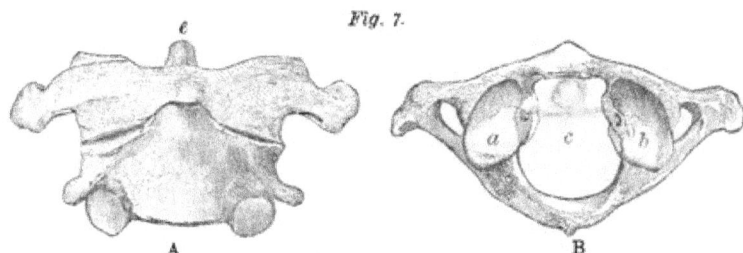

Fig. 7.

B, *the first cervical vertebra, the atlas ;* A, *the atlas, and the second cervical vertebra,
the axis ;* e, *the odontoid process ;* c, *the foramen.*

The Skull Articulates with (is jointed to) the spine
in a peculiar manner. On the top of the upper vertebra (atlas *) are two little hollows (*a*, *b*, Fig. 7),
nicely packed and lined with the synovial membrane, into which fit the corresponding projections
on the lower part of the skull, and thus the head can
rock to and fro. The second vertebra (axis) has a
peg, *e*, which projects through a hole, *c*, in the first.

The surfaces of both vertebræ are so smooth that
they easily glide on each other, and thus, when we
move the head sidewise, the atlas turns around the
peg, *e*, of the axis.

The Ribs, also twenty-four in number, are arranged
in pairs on each side of the chest. At the back, they

---

* Thus called because, as, in ancient fable, the god Atlas supported the world on
his shoulders, so in the body this bone bears the head.

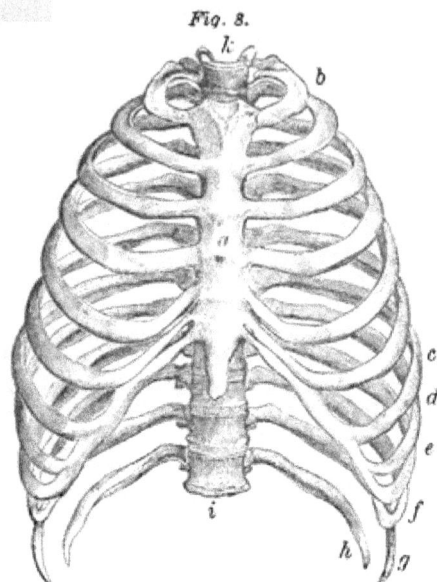

Fig. 8.

The Thorax, or Chest : a, the sternum ; b to c, the true ribs ; d to b, the false ribs ; g, h, the floating ribs ; i k, the dorsal vertebræ.

are all attached to the spine. In front, the upper seven pairs are tied by cartilages to the breast-bone (sternum); three are fastened to each other and the cartilage above, and two, the floating ribs, are loose.

The natural form of the chest is that of a cone diminishing upward. But, owing to the tightness of the clothing commonly worn, the reverse is often the case. The long, slender ribs give lightness,* the arched form confers strength, and the cartilages impart elasticity,—properties essential to the protection of the delicate organs within, and to freedom of motion in respiration. (See note, p. 80.)

* If the chest-wall were in one bone thick enough to resist a blow. it would be unwieldy and heavy. As it is, the separate bones bound by cartilages yield gradually, and diffuse the force among them all, and so are rarely broken.

Fig. 9.

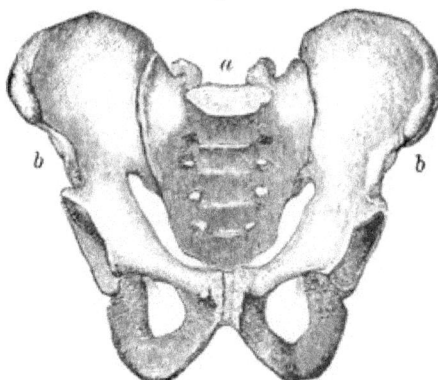

The Pelvis : a, the sacrum ; b, b, the right and the left innominatum.

The Hip-bones, called by anatomists the innominata, or nameless bones, form an irregular basin
styled the *pelvis* (*pelvis,* a basin). In the upper part,
is the foot of the spinal column—a wedge-shaped
bone termed the *sacrum** (sacred), firmly planted
here between the wide-spreading and solid bones of
the pelvis, like the keystone to an arch, and giving
a steady support to the heavy burden above.

3. THE LIMBS.

Two Sets of Limbs branch from the trunk, viz.:
the upper, and the lower. They closely resemble each
other. The arm corresponds to the thigh ; the fore-
arm, to the leg ; the wrist, to the ankle ; the fingers,
to the toes. The fingers and the toes are so much
alike that they receive the same name, *digits,* while
the several bones of both have also the common
appellation, *phalanges.* The differences which exist

---

* So called because it was anciently offered in sacrifice.

grow out of their varying uses. The foot is characterized by strength ; the hand, by mobility.

1. The Upper Limbs.—The Shoulder.—The bones of the shoulder are the collar-bone (clavicle), and the shoulder-blade (scapula). The *clavicle* (*clavis*, a key) is a long, slender bone, shaped like the Italic *f*. It is fastened at one end to the breast-bone and the first rib, and, at the other, to the shoulder-blade. (See Fig. 1.) It thus holds the shoulder-joint out from the chest, and gives the arm greater play. If it be removed or broken, the head of the arm-bone will fall, and the motions of the arm be greatly restricted.*

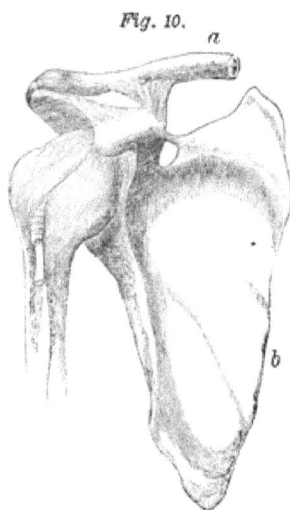

Fig. 10.

The Shoulder-joint ; a, the clavicle ; b, the scapula.

The Shoulder-blade is a thin, flat, triangular bone, fitted to the top and back of the chest, and designed to give a foundation for the muscles of the shoulder.

The Shoulder-joint. — The arm-bone, or *humerus*, articulates with the shoulder-blade by a ball-and-socket joint. This consists of a cup-like cavity in the latter bone, and a rounded head in the former, to fit it,—thus affording a free rotary motion. The shallowness of the socket accounts for the frequent dislocation of this joint, but a deeper one would diminish the easy swing of the arm.

---

* Animals which use the forelegs only for support (as the horse, ox, etc) do not possess this bone. "It is found in those that dig, fly, climb, and seize."

Fig. 11.

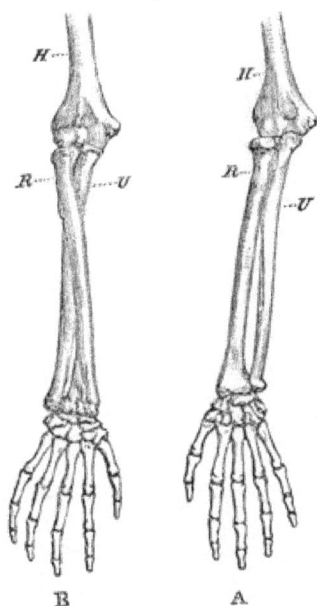

B   A

Bones of the right Fore-arm; H,
the humerus; R. the radius; and
U, *the ulna.*

The Elbow. — At the elbow, the humerus articulates with the *ulna* — a slender bone on the inner side of the forearm — by a hinge-joint which admits of motion in only two directions, i. e., backward and forward. The ulna is small at its lower end ; the *radius,* or large bone of the forearm, on the contrary, is small at its upper end, while it is large at its lower end, where it forms the wrist-joint. At the elbow, the head of the radius is convex and fits into a shallow cavity in the ulna, while at the wrist the ulna plays in a similar socket in the radius. Thus the radius may roll over and even cross the ulna.

The Wrist, or *carpus,* consists of two rows of very irregular bones, one of which articulates with the fore-arm ; the other, with the hand. They are placed side to side and so firmly fastened as to admit of only a gliding motion. This gives little play, but great strength, elasticity, and power of resisting shocks.

The Hand.—The *metacarpal* (*meta,* beyond ; and *karpos,* wrist), or bones of the palm, support each a thumb or finger. Each finger has three bones while the thumb has only two. The first bone of the thumb, standing apart from the rest, enjoys a special freedom of motion, and adds greatly to the usefulness of the hand.

Fig. 12.

The first bone (Figs. 11, 12) of each finger is so attached to the corresponding metacarpal bone as to move in several directions upon it, but the other phalanges form hinge-joints.

The fingers are named in order : the thumb, the

Bones of the Hand and the Wrist.

index, the middle, the ring, and the little finger. Their different lengths cause them to fit the hollow of the hand when it is closed, and probably enable us more easily to grasp objects of varying size. If the hand clasps a ball, the tips of the fingers will be in a straight line.

The hand in its perfection belongs only to man. Its elegance of outline, delicacy of mold, and beauty of color have made it the study of artists ; while its exquisite mobility and adaptation as a perfect instrument have led many philosophers to attribute man's superiority even more to the hand than to the mind.*

* How constantly the hand aids us in explaining or enforcing a thought ! We affirm a fact by placing the hand as if we would rest it firmly on a body ; we deny by a gesture putting the false or erroneous proposition away from us ; we express doubt by holding the hand suspended, as if hesitating whether to take or reject. When we part from dear friends, or greet them again after long absence, the hand

Fig. 13.

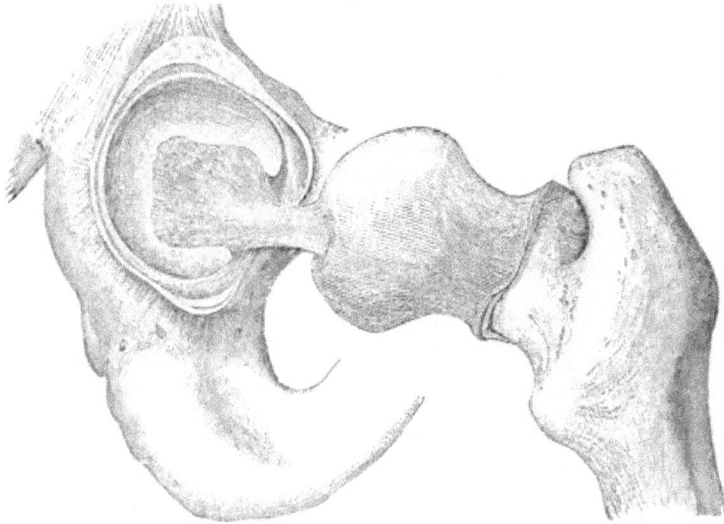

The Hip-joint.

2. The Lower Limbs.—The Hip.

The thigh-bone, or *femur*, is the largest and necessarily the strongest in the skeleton, since at every step it has to bear the weight of the whole body. It articulates with the hip-bone by a ball-and-socket joint. Unlike the shoulder-joint, the cup here is deep, thus affording less play, but greater strength. It fits so tightly that the pressure of the air largely aids in keeping the bones in place.* Indeed, when the muscles are cut away, great force is required to detach the limbs.

extends toward them as if to retain, or to bring them sooner to us. If a recital or a proposition is revolting, we reject it energetically in gesture as in thought. In a friendly adieu we wave our good wishes to him who is their object; but when it expresses enmity, by a brusque movement we sever every tie. The open hand is carried backward to express fear or horror, as well as to avoid contact; it goes forward to meet the hand of friendship; it is raised suppliantly in prayer toward Him from whom we hope for help: it caresses lovingly the downy cheek of the infant, and rests on its head invoking the blessing of Heaven.— *Wonders of the Human Body.*

* In order to test this, a hole was bored through a hip-bone so as to admit air into the socket; the thigh-bone at once fell out as far as the ligaments would permit. An

The Knee is strengthened by the *patella*, or knee-pan (*patella*, little dish), a chestnut-shaped bone firmly fastened over the joint.

The shin-bone, or *tibia*, the large, triangular bone on the inner side of the leg, articulates both with the femur and the foot by a hinge-joint. The knee-joint is so made, however, as to admit of a slight rotary motion when the limb is not extended.

The *fibula* (*fibula*, a clasp), the small, outside bone of the leg, is firmly bound at both ends to the tibia. (See Fig. 1.) It is immovable, and, as the tibia bears the principal weight of the body, the chief use of this second bone seems to be to give more surface to which the muscles may be attached.*

The Foot.—The general arrangement of the foot is strikingly like that of the hand (Fig. 1). The several parts are the *tarsus*, the *metatarsus*, and the *phalanges*. The graceful arch of the foot, and the numerous bones joined by cartilages, give an elasticity to the step that could never be attained by a single, flat bone. The toes naturally lie straight forward in the line of the foot. Few persons in civilized nations, however, have naturally-formed feet. The big toe

experiment was also devised whereby a suitably-prepared hip-joint was placed under the receiver of an air-pump. On exhausting the air, the weight of the femur caused it to drop out of the socket, while the re-admission of the air raised it to its place. Without this arrangement, the adjacent muscles would have been compelled to bear the additional weight of the thigh-bone every time it was raised. Now the pressure of the air rids them of this unnecessary burden, and hence they are less easily fatigued.—*Weber.*

* A young man in the hospital at Limoges had lost the middle part of his tibia. The lost bone was not reproduced, but the fibula, the naturally weak and slender part of the leg, became thick and strong enough to support the whole body. An experiment has been performed which illustrates this idea still more strikingly. An inch of the middle part of the fibula of an animal was cut out. A long time afterward the beast was killed, when the tibia was found to have become considerably larger in that part of it which corresponded exactly with the defect in the fibula.—*Stanley's Lectures.*

is crowded upon the others, while crossed toes, nails grown-in, enormous joints, corns, and bunions abound.

The Cause of these Deformities is found in the shape and size of fashionable boots and shoes. The sole ought to be large enough for full play of motion, the uppers should not crowd the toes, and the heels should be low, flat, and broad. As it is, there is a constant warfare between Nature and our shoe-makers,* and we are the victims. The narrow point in front pinches our toes, and compels them to over-ride one another; the narrow sole compresses the arch; while the high heel, by throwing all the weight forward on the toes, strains the ankle, and, by sending the pressure where Nature did not design it to fall, causes that joint to become en-larged. The body bends forward to meet the demand of this new motion, and thus loses its up-rightness and beauty, making our gait stiff and un-graceful.

Diseases, etc.—1. The Rickets are caused by a lack of mineral matter in the bones, rendering them soft and pliable, so that they bend under the weight of the body. They thus become permanently dis-torted, and of course are weaker than if they were straight.† The disease is cured by a more nutritive

---

* When we are measured for boots or shoes, we should *stand* on a sheet of paper, and have the shoemaker mark with a pencil the exact outline of our feet as they bear our whole weight. When the shoe is made, the sole should exactly cover this outline.

† Just here appears an exceedingly beautiful provision. As soon as the dispro-portion of animal matter ceases, a larger supply of mineral is sent to the weak points, and the bones actually become thicker, denser, harder, and consequently stronger at the very concave part where the stress of pressure is greatest.—*Watson's Lectures.* We shall often have occasion to refer to similar wise and providential arrangements whereby the body is enabled to remedy defects, and to prepare for accidents.

diet, or by taking phosphate of lime to supply the lack.

2. A FELON is a swelling of the finger or thumb, usually of the last joint. It is marked by an accumulation beneath the periosteum and next the bone. The physician will merely cut through the periosteum, and let out the effete matter.

3. BOWLEGS are caused by children standing on their feet before the bones of the lower limbs are strong enough to bear their weight. The custom of encouraging young children to stand up by means of a chair or the support of the hand, while the bones are yet soft and pliable, is a cruel one, and liable to produce permanent deformity. Nature will set the child on its feet when the proper time comes.

4. CURVATURE OF THE SPINE.—When the spine is bent, the packing between the vertebræ becomes compressed on one side into a wedge-like shape. After a time, it will lose its elasticity, and the spine become distorted. This occurs frequently in the case of students who bend forward to bring their eyes nearer their books, instead of lifting their books nearer their eyes, or who raise their right shoulder above their left when writing at a desk which is too high. Round shoulders, small, weak lungs, and, oftentimes, diseases of the spine are the consequences. An erect posture in reading or writing conduces not alone to beauty of form, but also to health of body.

5. SPRAINS are produced when the ligaments which bind the bones of a joint are strained, twisted, or torn from their attachments. They are quite as harmful as a broken bone, and require careful atten-

tion lest they lead to a crippling for life. The use of a sprained limb may permanently impair its usefulness. Hence, the joint should be kept quiet, even after the immediate pain is gone.

6. A DISLOCATION is produced by the rupture of the tissues of the joint so that the head of the bone is driven out of its socket and into some other place both by the force of the blow which caused the injury and by the contraction of the muscles.

PRACTICAL QUESTIONS.

1. Why does not a fall hurt a child as much as it does a grown person ?

2. Should a young child ever be urged to stand or walk ?

3. What is meant by " breaking one's neck " ?

4. Should chairs or benches have straight backs ?

5. Should a child's feet be allowed to dangle from a high seat ?

6. Why can we tell whether a fowl is young by pressing on the point of the breast-bone ?

7. What is the use of the marrow in the bones ?

8. Why is the shoulder so often put out of joint ?

9. How can you tie a knot in a bone ?

10. Why are high pillows injurious ?

11. Is the " Grecian bend " a healthful position ?

12. Should a boot have a heel-piece ?

13. Why should one always sit and walk erect ?

14. Why does a young child creep rather than walk ?

15. What is the natural direction of the big toe ?

II.

THE MUSCLES.

" *Behold the outward moving frame,*
Its living marbles jointed strong
With glistening band and silvery thong,
And link'd to reason's guiding reins
By myriad rings in trembling chains,
Each graven with the threaded zone
Which claims it as the Master's own."

HOLMES.

THE MUSCLES.

1. THE USE, STRUCTURE, AND ACTION OF THE MUSCLES.
1. The use of the muscles.
2. Contractility of the muscles.
3. Arrangement of the muscles.
4. The two kinds of muscles.
5. The structure of the muscles.
6. The tendons for fastening muscles.
7. The muscles and bones as levers.
8. The effect of big joints.
9. Action of the muscles in standing.
10. Action of the muscles in walking.

2. THE MUSCULAR SENSE.

3. HYGIENE OF THE MUSCLES.
1. Necessity of Exercise.
2. Time for Exercise.
3. Kinds of Exercise.

4. WONDERS OF THE MUSCLES.

5. DISEASES..........
1. St. Vitus's Dance.
2. Convulsions.
3. Locked-jaw.
4. Gout.
5. Rheumatism.
6. Lumbago.
7. A Ganglion.

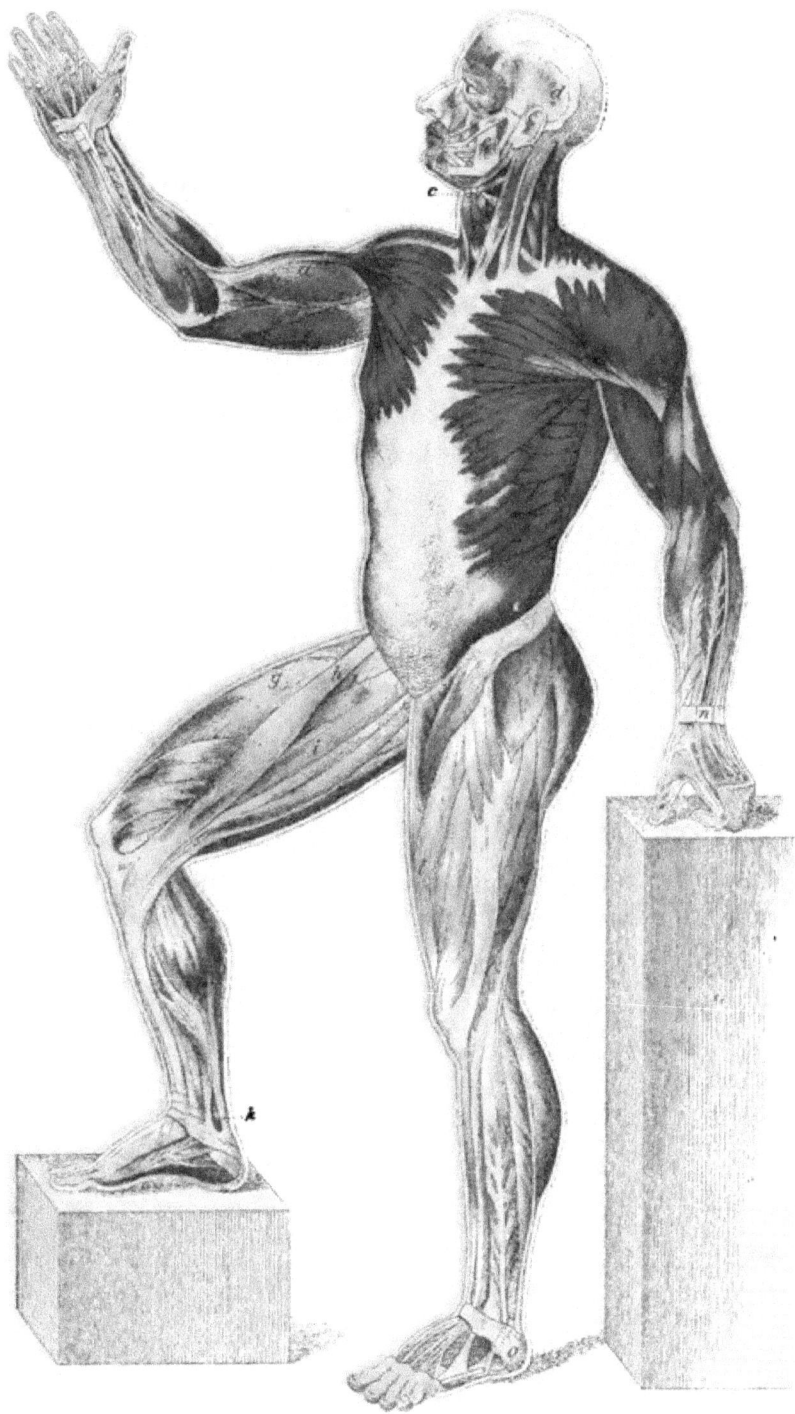

THE MUSCLES.

THE **Use of the Muscles.**—The skeleton is the image of death. Its unsightly appearance instinctively repels us. We have seen, however, what uses it subserves in the body, and how the ugly-looking bones abound in nice contrivances and ingenious workmanship. In life, the framework is hidden by the flesh. This covering is a mass of muscles, which not only give form and symmetry to the body, but also produce its varied movements.

In Fig. 14, we see the large exterior muscles. Beneath these are many others ; while deeply hidden within are tiny, delicate ones, too small to be seen with the naked eye. There are, in all, about five hundred, each having its special use, and all working in exquisite harmony and perfection.

Contractility.—The peculiar property of the muscles is their power of contraction, whereby they decrease in length and increase in thickness.* This may be caused by an effort of the will, by cold, by a sharp blow, &c. It does not cease at death, but, in certain cold-blooded animals, a contraction of the muscles is often noticed long after the head has been cut off.

---

* The maximum force of this contraction has been estimated as high as from 85 to 114 pounds per square inch.

Arrangement of the Muscles.* — The muscles are nearly all arranged in pairs, each with its antagonist, so that, as they contract and expand alternately, the bone to which they are attached is moved to and fro.

If you grasp the arm tightly with your hand just above the elbow-joint, and bend the forearm, you will feel the muscle on the inside (biceps, *a*, Fig. 14) swell, and become hard and prominent, while the outside muscle (triceps, *f*) will be relaxed. Now straighten the arm, and the swelling and hardness of the inside muscle will vanish, while the outside one will, in turn, become rigid. So, also, if you clasp the arm just below the elbow, and then open and shut the fingers, you can feel the alternate expanding and relaxing of the muscles on opposite sides of the arms.

If the muscles on one side of the face become palsied, those on the other side will draw the mouth that way. Squinting is caused by one of the straight muscles of the eye (Fig. 17) contracting more strongly than its antagonist.

Kinds of Muscles.—There are two kinds of muscles, the *voluntary*, which are under the control of our will, and the *involuntary*, which are not. Thus our limbs stiffen or relax as we please, but the heart beats on by day and by night. The eyelid, however,

---

* " Could we behold properly the muscular fibers in operation, nothing, as a mere mechanical exhibition, can be conceived more superb than the intricate and combined actions that must take place during our most common movements. Look at a person running or leaping, or watch the motions of the eye. How rapid, how delicate, how complicated, and yet how accurate, are the motions required! Think of the endurance of such a muscle as the heart, that can contract, with a force equal to sixty pounds, seventy-five times every minute, for eighty years together, without being weary."

is both voluntary and involuntary, so that while we wink constantly without effort, we can, to a certain extent, restrain or control the motion.

Structure of the Muscles.—If we take a piece of lean beef and wash out the red color, we can easily detect the fine fibers of which the meat is composed. In boiling corned beef for the table, the fibers often separate, owing to the dissolving of the delicate tissue which bound them together. By means of the microscope, we find that these fibers are made up of minute filaments (*fibrils*), and that each fibril is composed of a row of small cells arranged like a string of beads. This gives the muscles a peculiar striped (striated) appearance.* The cells are filled with a fluid or semi-fluid mass of living (protoplasmic) matter.

The binding of so many threads into one bundle † confers great strength, according to a mechanical

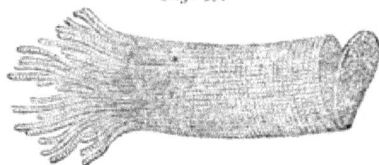

Fig. 15.

Microscopic view of a Muscle, showing, at one end, the fibrillæ ; and, at the other, the disks, or cells, of the fiber.

principle that we see exemplified in suspension bridges, where the weight is sustained, not by bars

---

* The involuntary muscles consist generally of smooth, fibrous tissue, and form sheets or membranes in the walls of hollow organs. By their contraction they change the size of cavities which they enclose. Some functions, however, like the action of the heart, or the movements of deglutition (swallowing), require the rapid, vigorous contraction, characteristic of the voluntary muscular tissue.—*Flint.*

† We shall learn hereafter how these fibers are firmly tied together by a mesh of fine connective tissue which dissolves in boiling, as just described.

of iron, but by small wires twisted into massive ropes.

The Tendons.—The ends of the muscles are generally attached to the bone by strong, flexible, but inelastic tendons.* The muscular fibers spring from the sides of the tendon, so that more of them can act upon the bone than if they went directly to it. Besides, the small, insensible tendon can better bear the exposure of passing over a joint, and be more easily lodged in some protecting groove, than the broad, sensitive muscle. This mode of attachment gives to the limbs strength, and elégance of form. Thus, for example, if the large muscles of the arm extended to the hand, they would make it bulky and clumsy. The tendons, however, reach only to the wrist, whence fine cords pass to the fingers (Fig. 16).

Fig. 16.

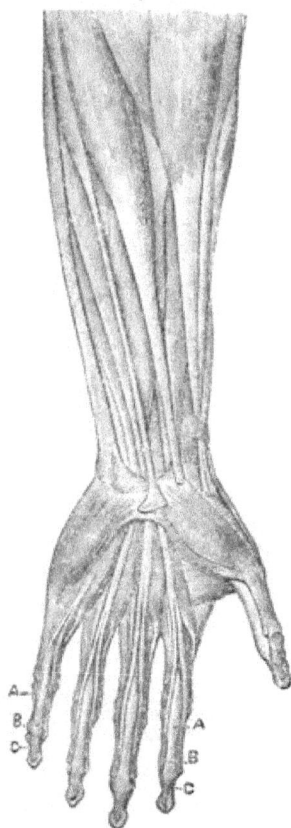

Tendons of the Hand

Here we notice two other admirable arrangements. 1. If the long tendons at the wrist on contracting should rise, projections would be made and thus the beauty of

* The tendons may be easily seen in the leg of a turkey as it comes on our table; so, at a Thanksgiving dinner, we may study Physiology while we pick the bones.

the slender joint be marred. To prevent this, a stout band or bracelet of ligament holds them down to their place. 2. In order to allow the tendon which moves the last joint of the finger to pass through, the tendon which moves the second joint divides at its attachment to the bone (Fig. 16). This is the most economical mode of packing the muscles, as any other practicable arrangement would increase the bulk of the slender finger.

Fig. 17.

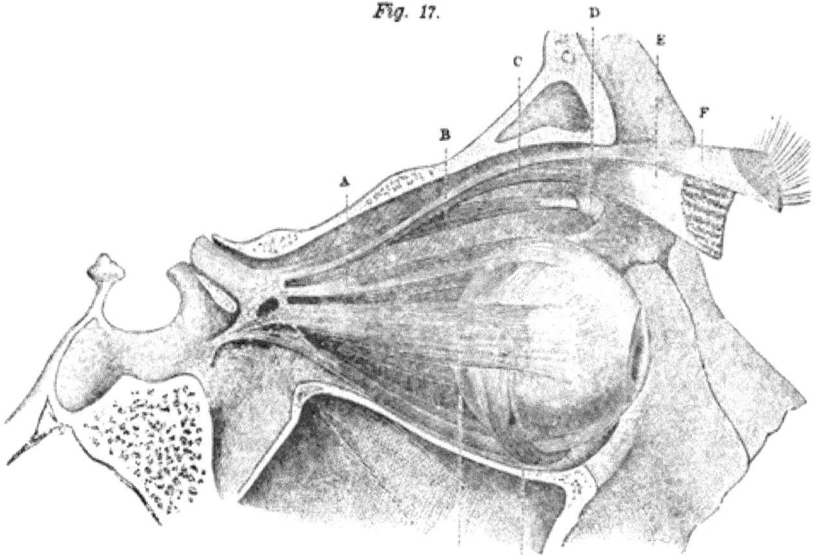

The Muscles of the Right Eye. A. *superior straight*; B. *superior oblique passing through a pulley,* D ; G. *inferior oblique ;* H, *external straight, and, back of it, the internal straight muscle.*

Since the tendon cannot always pull in the direction of the desired motion, some contrivance is necessary to meet the want. The tendon (B) belonging to one of the muscles of the eye, for example, passes through a ring of cartilage, and thus a rotary motion is secured.

The Levers of the Body.*—In producing the motions of the body, the muscles use the bones as levers.

Fig. 18.

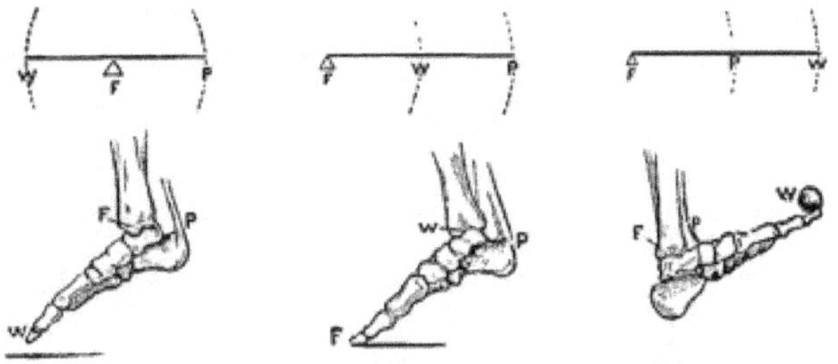

The three classes of Levers, and also the foot as a Lever.

We see an illustration of the *first class* of levers in the movements of the head. The back or front of the head is the weight to be lifted, the backbone is

Fig. 19.

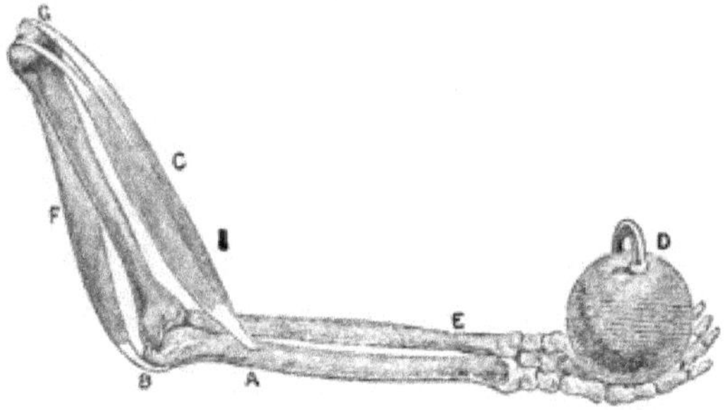

The hand as a Lever of the third class.

the fulcrum on which the lever turns, and the muscles at the back or front of the neck exert the power by which we toss or bow the head.

* A *lever* is a stiff bar resting on a point of support, called the *fulcrum* (F), and having connected with it a *weight* (W) to be lifted, and a *power* (P) to move it. There are three classes of levers according to the arrangement of the power, weight, and fulcrum. In the 1st class, the F is between the P and W; in the 2d, the W is between the P and F; and in the 3d, the P is between the W and F (Fig. 18). A pump-handle is an example of the first; a lemon-squeezer, of the second; and a pair of fire-tongs, of the third. See *Physics*, pp. 69—71, for a full description of this subject, and many illustrations.

When we raise the body on tiptoe, we have an instance of the *second class*. Here, our toes resting on the ground form the fulcrum, the muscles of the calf (gas-troc-né-mi-us, *j*, and so-lé-us, Fig. 14), acting through the tendon of the heel,* are the power, and the weight is borne by the ankle joint.

An illustration of the *third class* is found in lifting the hand from the elbow. The hand is the weight, the elbow the fulcrum, and the power is applied by the biceps muscle at its attachment to the radius. (A, Fig. 19.) In this form of the lever there is a great loss of force, because it is applied at such a distance from the weight, but there is a gain of velocity, since the hand moves so far by such a slight contraction of the muscle. The hand is required to perform quick motions, and therefore this mode of attachment is wisely adopted.

The nearer the power is applied to the resistance, the more easily the work is done. In the lower jaw, for example, the jaw is the weight, the fulcrum is the hinge-joint at the back, and the muscles (temporal, *d*, and the mas'-se-ter, *e*, Fig. 14) on each side are the power.† They act much closer to the resist-

---

* This is called the Tendon of Achilles (*k*, Fig. 14), and is so named because, as the fable runs, when Achilles was an infant his mother held him by the heel while she dipped him in the River Styx, whose water had the power of rendering one invulnerable to any weapon. His heel, not being wet, was therefore his weak point, and here Paris, at last, directed the fatal arrow which killed the famous hero.—" This tendon will bear 1000 lbs. weight before it will break."—*Mapother*.

The horse is said to be " hamstrung " and so rendered useless, when the Tendon of Achilles is cut.

† We may feel the contraction of the masseter by placing our hand on the face when we work the jaw, while the temporal can be readily detected by putting the fingers on the temple while we are chewing. The tendon of the muscle (digastric) —one of those which open the jaw—passes through a pulley (*c*, Fig. 14) somewhat like the one in the eye.

ance than those in the hand, since here we desire
force, and there speed.

Fig. 20.

The Knee-joint:
b, the patella;
f, the tendon.

The Enlargement of the Bones at the
Joints not only affords greater surface
for the attachment of the muscles, as
we have seen, but also enables them
to work to better advantage. Thus, in
Fig. 20 it is evident that a muscle acting in the line *fb* would not bend the
lower limb so easily as if it were acting
in the line *fh*, since in the former case
its force would be about all spent in
drawing the bones more closely together, while in the latter it would pull
more nearly at a right angle. Thus the tendon *f*,
by passing over the patella, which is itself pushed
out by the protuberance *b* of the thigh-bone, pulls
at a larger angle,* and so the leg is thrown forward with ease in walking and with great force in
kicking.

How We Stand Erect.—The joints play so easily,
and the center of gravity in the body is so far above
the foot, that the skeleton cannot of itself hold our
bodies upright. Thus it requires the action of many
muscles to maintain this position. The head so
rests upon the spine as to tend to fall in front, but
the muscles of the neck steady it in its place.† The

---

* The chief use of the processes of the spine (Fig. 6) and other bones is, in the
same way, to throw out the point on which the power acts as far from the fulcrum
as possible. The projections of the ulna ("funny-bone") behind the elbow, and
that of the heel-bone to which the Tendon of Achilles is attached, are excellent
illustrations (Fig. 1).

† In animals the jaws are so heavy, and the place where the head and spine join
is so far back, that there can be no balance as there is in man. There are therefore

hips incline forward, but are held erect
by the strong muscles of the back. The
trunk is nicely balanced on the head of
the thigh-bones. The great muscles of
the thigh acting over the knee-pan tend
to bend the body forward, but the muscles of the calf neutralize this action.
The ankle, the knee, and the hip lie
in nearly the same line, and thus the
weight of the body rests directly on the
key-stone of the arch of the foot. So
perfectly do these muscles act that we
never think of them until science calls
our attention to the subject, and yet to
acquire the necessary skill to use them
in our infancy needed patient lessons,
much time, and many hard knocks.

How We Walk.—Walking is as complex an act as standing. It is really a
perilous performance, which has become safe only because of constant
practice. We see how violent it is when
we run against a post in the dark, and
find with what headlong force we were
hurling ourselves forward. Holmes has

Fig. 21.

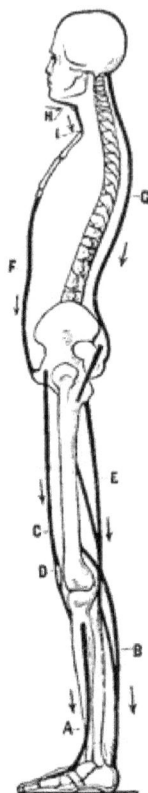

Action of the Muscles which keep the body erect.

well defined walking as a perpetual falling with a
constant self-recovery. Standing on one foot we
let the body fall forward, and swing the other leg
ahead like a pendulum. Planting that foot on the

large muscles in their necks. We readily find that we have none if we get on "all
fours" and try to hold up the head. On the other hand, gorillas and apes cannot
stand up erect like man. Their head, trunk, legs, etc., are not balanced by muscles,
so as to be in line with one another.

ground, to save the body from falling further, we then swing the first foot forward again to repeat the same operation.*

The shorter the pendulum, the more rapidly it vibrates ; and so short-legged people take quicker and shorter steps than long-legged ones.† We are shorter when walking than when standing still, because of this falling forward to take a step in advance.‡

In running, we incline the body more, and so, as it were, fall faster. When we walk, one foot is on the ground all the time, and there is an instant when both feet are planted upon it ; but in running there is an interval of time in each step when both feet are off the ground, and the body is wholly unsupported. As we step alternately with the feet, we are inclined to turn the body first to one side and then to the other. This movement is sometimes counterbalanced by swinging the hand on the opposite side. §

* It is a curious fact that one side of the body tends to out-walk the other ; and so, when a man is lost in the woods, he often goes in a circle, and at last comes round to the spot whence he started.

† In this respect, Tom Thumb was to Magrath, whose skeleton, eight and one-half feet high, is now in the Dublin Museum, what a little, fast-ticking, French mantel-clock is to a big, old-fashioned, upright, corner time-piece.

‡ Women find that a gown that will swing clear of the ground when they are standing still, will drag the street when they are walking. The length of the step may be increased by muscular effort as when a line of soldiers keep step in spite of their having legs of different lengths. Such a mode of walking is necessarily fatiguing.

§ "In ordinary walking the speed is nearly four miles an hour, and can be kept up for a long period. But exercise and a special aptitude for it enable some men to walk great distances in a relatively short space of time. Trained walkers have gone seventy-five miles in twenty hours, and walked the distance of thirty-seven miles at the rate of five miles an hour. The mountaineers of the Alps are generally good walkers, and some of them are not less remarkable for endurance than for speed. Jacques Balmat, who was the first to reach the summit of Mont Blanc, at sixteen years of age could walk from the hamlet of the Pèlerins to the mountain of La Côte

The Muscular Sense.—When we lift an object, we feel a sensation of weight, which we can compare with that experienced in lifting another body.* By care we may cultivate this sense so as to form a very precise estimate of the weight of a body by balancing it in the hand. The muscular sense is useful to us in many ways. It guides us in standing or moving. We gratify it when we walk erect and with an elastic step, and by dancing, jumping, skating, and gymnastic exercises.

Necessity of Exercise.—The effect of exercise upon a muscle is very marked.† By use it grows larger, and becomes hard, compact, and darker-colored ; by disuse it decreases in size, and becomes soft, flabby, and pale.

Violent exercise, however, is injurious, since we

in two hours,—a distance which the best trained travelers required from five to six hours to get over. At the time of his last attempt to reach the top of Mont Blanc, this same guide, then twenty years old, passed six days and four nights without sleeping or reposing a single moment. One of his sons, Edward Balmat, left Paris to join his regiment at Genoa ; he reached Chamonix the fifth day at evening, having walked 340 miles. After resting two days, he set off again for Genoa, where he arrived in two days. Several years afterward, this same man left the baths at Louêche at two o'clock in the morning, and reached Chamonix at nine in the evening, having walked a distance equal to about seventy-five miles in nineteen hours. In 1844, an old guide of De Saussure, eighty years old, left the hamlet of Prats. in the valley of Chamonix, in the afternoon, and reached the Grand Mulets at ten in the evening ; then, after resting some hours. he climbed the glacier to the vicinity of the Grand Plateau, which has an altitude of about 13,000 feet, and then returned to his village without stopping."— *Wonders of the Body.*

* If a small ivory ball be allowed to roll down the cheek toward the lips, it will appear to increase in weight. In general, the more sensitive parts of the body recognize smaller differences in weight, and the right hand is more accurate than the left. We are very apt, however, to judge of the weight of a body from previous conceptions. Thus, shortly after Sir Humphrey Davy discovered the metal potassium, he placed a piece of it in Dr. Pierson's hand, who exclaimed, " Bless me ! How heavy it is ! " Really, however, potassium is so light that it will float on water like cork.

† The greater size of the breast (pectoral muscle) of a pigeon, as compared with that of a duck, shows how muscle increases with use. The breast of a chicken is white because it is not used for flight, and therefore gets little blood.

then tear down faster than nature can build up. Feats of strength are not only hurtful, but dangerous. Often the muscles are strained or ruptured, and blood-vessels burst in the effort to outdo one's companions.*

Two thousand years ago, Isocrates, the Greek rhetorician, said, "Exercise for health, not for strength." The cultivation of muscle for its own sake is a return to barbarism, while it enfeebles the mind, and ultimately the body. The ancient gymnasts are said to have become prematurely old, and the trained performers of our own day soon suffer from the strain they put upon their muscular system. Few men have vigor sufficient to become both athletes and scholars. Exercise should, therefore, merely supplement the deficiency of our usual employment. *A sedentary life needs daily, moderate exercise, which always stops short of fatigue.* This is a law of health.

No education is complete which fails to provide for the development of the muscles. Recesses should be as strictly devoted to play as study-hours are to work. Were gymnastics or calisthenics as regular an exercise as grammar or arithmetic, fewer pupils would be compelled to leave school on account of ill health ; while spinal curvatures, weak backs, and ungraceful gaits would no longer characterize so many of our best institutions.

Time for Exercise.—We should not exercise after long abstinence from food, nor immediately after a

---

* Instances have been known of children falling dead from having carried to excess so pleasant and healthful an amusement as jumping the rope, and of persons rupturing the Tendon of Achilles in dancing.

meal, unless the meal or the exercise be very light. There is an old-fashioned prejudice in favor of exercise before breakfast—an hour suited to the strong and healthy, but entirely unfitted to the weak and delicate. On first rising in the morning, the pulse is low, the skin relaxed, and the system susceptible to cold. Feeble persons, therefore, need to be braced with food before they brave the out-door air.

What Kind of Exercise to Take.—For children, games are unequalled. Walking, the universal exercise,* is beneficial, as it takes one into the open air and sunlight. Running is better, since it employs more muscles, but must not be pushed to excess, as it taxes the heart, and may lead to disease of that organ. Rowing is more effectual in its general development of the system. Swimming employs the muscles of the whole body, and is a valuable acquirement, as it may be the means of saving life. Horseback riding is a fine accomplishment, and refreshes mind and body alike. Gymnastic or calisthenic exercises, when carefully selected, and not indulged immoderately, bring into play all the muscles of the body, and become preferable to any other mode of in-door exercise.†

* The custom of walking, so prevalent in England, has doubtless much to do with the superior physique of its people. It is considered nothing for a woman to take a walk of eight or ten miles, and long pedestrian excursions are made to all parts of the country. The benefits which accrue from such an open-air life are sadly needed by the women of our own land. A walk of half-a-dozen miles should be a pleasant recreation for any healthy person.

† The employment of the muscles in exercise not only benefits their especial structure, but it acts on the whole system. When the muscles are put in action, the capillary blood-vessels with which they are supplied become more rapidly charged with blood, and active changes take place, not only in the muscles, but in all the surrounding tissues. The heart is required to supply more blood, and accordingly beats more rapidly in order to meet the demand. A larger quantity of blood is sent through the lungs, and larger supplies of oxygen are taken in and carried to the

The Wonders of the Muscles.—The grace, ease, and rapidity with which the muscles contract are astonishing. By practice, they acquire a facility which we call mechanical. The voice may utter 1500 letters in a minute, yet each requires a distinct position of the vocal organs. We train the muscles of the fingers till they glide over the keys of the piano, executing the most exquisite and difficult harmony. In writing, each letter is formed by its peculiar motions, yet we make them so unconsciously that a skilful penman will describe beautiful curves while thinking only of the idea that the sentence is to express. The mind of the violinist is upon the music which his right hand is executing, while his left determines the length of the string and the character of each note so carefully that not a false sound is heard, although the variation of a hair's breadth would cause a discord. In the arm of a blacksmith, the biceps muscle may grow into the solidity almost of a club ; the hand of a prize-fighter will strike a blow like a sledge-hammer : while the engraver traces lines invisible to the naked eye,

various tissues. The oxygen. by combining with the carbon of the blood and the tissues, engenders a larger quantity of heat. which produces an action on the skin. in order that the superfluous warmth may be disposed of. The skin is thus exercised, as it were, and the sudoriparous and sebaceous glands are set at work. The lungs and skin are brought into operation, and the lungs throw off large quantities of carbonic acid, and the skin large quantities of water, containing in solution matters which, if retained, would produce disease in the body. Wherever the blood is sent, changes of a healthful character occur. The brain and the rest of the nervous system are invigorated, the stomach has its powers of digestion improved, and the liver, pancreas, and other organs perform their functions with more vigor. By want of exercise. the constituents of the food which pass into the blood are not oxidized, and products which produce disease are engendered. The introduction of fresh supplies of oxygen induced by exercise oxidizes these products, and renders them harmless; all other things being the same, it may be laid down as a rule that those who take the most exercise in the open air will live the longest.—*Lankester.*

and the fingers of the blind acquire a delicacy that almost supplies the place of the missing sense.

Diseases, etc. — 1. St. Vitus's Dance is a disease of the voluntary muscles, whereby they are in frequent, irregular, and spasmodic motion beyond the control of the will. All causes of excitement, and especially of fear, should be avoided, and the general health of the patient invigorated, as this disease is closely connected with a derangement of the nervous system.

2. Convulsions are an involuntary contraction of the muscles. Consciousness is wanting, while the limbs may be stiff or in spasmodic action. (See Appendix.)

3. Locked-jaw is a disease in which there are spasms and a contraction of the muscles, usually beginning in the lower jaw. It is serious, often fatal, yet it is sometimes caused by as trivial an injury as the stroke of a whip-lash, the lodgement of a bone in the throat, a fish-hook in the finger, or the puncture of the sole of the foot by a tack or a nail.

4. Gout is an acute pain located chiefly in the small joints of the foot, especially those of the great toe, which become swollen and extremely sensitive. It is generally brought on by high living.

5. Rheumatism affects mainly the connective, white, fibrous tissue of the larger joints. While gout is the punishment of the rich who live luxuriously and indolently, rheumatism afflicts the poor and the rich alike. There are two common forms of rheumatism—the inflammatory or acute, and the

chronic. The latter is of long continuance; the former terminates more speedily. The acute form is probably a disease of the blood, which carries with it some poisonous matter that is deposited where the fibrous tissue is most abundant.

The disease flies from one joint to another in the most unaccountable manner, and the pain caused by even the slightest motion deprives the sufferer of the use of the disabled part and its muscles. The chief danger to be feared is the possibility of its going to the heart. Any violent remedies, therefore, which would throw it from the surface are to be avoided. There is no generally-accepted mode of treating the disease. Warm fomentations are usually grateful. Chronic rheumatism—the result of repeated attacks of the acute—leads to great suffering, and oftentimes to disorganization of the joints, and an interference with the movements of the heart.

6. LUMBAGO is a rheumatic pain in the muscles of the small of the back.* It may be so moderate as to produce only a "lame back," or so severe as to disable, as in the case of a "crick in the back." Strong swimmers who sometimes suddenly drown without apparent cause are supposed to be seized in this way.

7. A GANGLION, or what is vulgarly called a

* Lumbago is really a form of myalgia, a disease which has its seat in the muscles, and may thus affect any part of the body. Doubtless much of what is commonly called "liver" or "kidney complaint" is only, in one case, myalgia of the chest or abdominal walls near the liver, or, in the other, of the back and loins near the kidneys. Chronic liver disease is comparatively rare in the northern States, and pain in the side is not a prominent symptom, while certain diseases of the kidneys, which are as surely fatal as pulmonary consumption, are not attended by pain in the back opposite these organs.— *Wey.*

"weak" or "weeping sinew," is a swelling of a bursa.* It sometimes becomes so distended by fluid as to be mistaken for bone. If on binding something hard upon it for a few days it does not disappear, a physician will remove the liquid by means of a hypodermic syringe, or perhaps "scatter" it by an external application of iodine.

PRACTICAL QUESTIONS.

1. What class of lever is the foot when we lift a weight on the toes?

2. Explain the movement of the body backward and forward, when resting upon the thigh-bone as a fulcrum.

3. What class of lever do we use when we lift the foot while sitting down?

4. Explain the swing of the arm from the shoulder.

5. What class of lever is used in bending our fingers?

6. What class of lever is our foot when we tap the ground with our toes?

7. What class of lever do we use when we raise ourselves from a stooping position?

8. What class of lever is the foot when we walk?

9. Why can we raise a heavier weight with our hand when lifting from the elbow than from the shoulder?

10. What class of lever do we employ when we are hopping, the thigh bone being bent up toward the body and not used?

11. Describe the motions of the bones when we are using a gimlet.

12. Why do we tire when we stand erect?

13. Why does it rest us to change our work?

---

* A bursa is a small sack containing a lubricating fluid to prevent friction where tendons play over hard surfaces. There is one shaped like an hour-glass on the wrist, just at the edge of the palm. By pressing back the liquid it contains, this bursa may be clearly seen.

14. Why and when is dancing a beneficial exercise?

15. Why can we exert greater force with the back teeth than with the front ones?

16. Why do we lean forward when we wish to rise from a chair?

17. Why does the projection of the heel-bone make walking easier?

18. Does a horse travel with less fatigue over a flat than a hilly country?

19. Can you move your upper jaw?

20. Are people naturally right or left-handed?

21. Why can so few persons move their ears by the muscles?

22. Is the blacksmith's right arm healthier than the left?

23. Boys often, though foolishly, thrust a pin into the flesh just above the knee. Why is it not painful?

24. Will ten-minutes practice in a gymnasium answer for a day's exercise?

25. Why would an elastic tendon be unfitted to transmit the motion of a muscle?

26. When one is struck violently on the head, why does he instantly fall?

27. What is the cause of the difference between light and dark meat in a fowl?

III.

THE SKIN.

— *A protection from the outer world, it is our only means of communicating with it. Insensible itself, it is the organ of touch. It feels the pressure of a hair, yet bears the weight of the body. It yields to every motion of that which it wraps and holds in place. It hides from view the delicate organs within, yet the faintest tint of a thought shines through, while the soul paints upon it, as on a canvas, the richest and rarest of colors.*

THE SKIN.

1. THE STRUCTURE OF THE SKIN.
1. The Cutis; its composition and character.
2. The Cuticle; its composition and character.
3. The value of the Cuticle.
4. The Complexion.

2. THE HAIR AND THE NAILS.
1. The Hair
  a. Description.
  b. Method of Growth.
  c. As an instrument of feeling.
  d. Indestructibility of the hair.
2. The Nails
  a. Uses.
  b. Method of growth.

3. THE MUCOUS MEMBRANE.
1. The Structure.
2. Connective Tissue.
3. Fat.

4. THE TEETH
1. Number and kinds of Teeth.
  1. The two sets
    1. The Milk Teeth.
    2. The Permanent Teeth
2. Structure of the Teeth.
3. The Setting of the Tooth in the Jaw.
4. The Decay of the Teeth.
5. The Preservation of the Teeth.

5. THE GLANDS
1. The two kinds
  1. Oil Glands.
  2. Perspiratory Glands.
2. The Perspiration.
3. The Absorbing Power of the Skin. (See Lymphatics.)

6. HYGIENE
1. About Washing and Bathing.
2. The Reaction.
3. Sea-Bathing.
4. Clothing
  a. General Principles.
  b. Linen.
  c. Cotton.
  d. Woolen.
  e. Flannel.
  f. Color of Clothing
  g. Structure of Clothing
  h. Insufficient Clothing.

7. DISEASES
1. Erysipelas.
2. Dropsy.
3. Corns.
4. In-growing Nails.
5 Warts.
6. Chilblains.
7. Wens.

THE SKIN.

THE **Skin** is a tough, thin, close-fitting garment for the protection of the tender flesh. Its perfect elasticity beautifully adapts it to every motion of the body. We shall learn hereafter that it is more than a mere covering, being an active organ, which does its part in the work of keeping in order the house in which we live. It oils itself to preserve its smoothness and delicacy, replaces itself as fast as it wears out, and is at once the perfection of use and beauty.

1. STRUCTURE OF THE SKIN.

Cutis and Cuticle.—What we commonly call the skin—viz., the part raised by a blister—is only the cuticle* or covering of the **cutis** or **true** skin. The latter is full of nerves and blood-vessels, while the former neither bleeds † nor gives rise to pain, neither suffers from heat nor feels the cold.

The cuticle is composed of small, flat cells or

---

* *Cuticula*, little skin. It is often styled the scarf-skin, and also the epidermis (*epi*, upon ; and *derma*, skin).

† We notice this in shaving : for if a razor goes below the cuticle it is followed by pain and blood. So insensible is this outer layer that we can run a pin through the thick mass at the roots of the nails without discomfort.

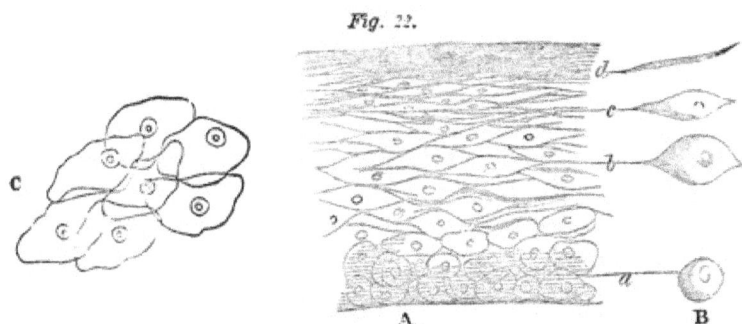

Fig. 22.

A represents a vertical section of the cuticle. B. lateral view of the cells. C. flat side of scales like d, magnified 250 diameters, showing the nucleated cells transformed into broad scales.

scales. These are constantly shed from the surface in the form of scurf, dandruff, etc., but are as constantly renewed from the cutis* below.

Under the microscope, we can see the round cells of the cuticle, and how they are flattened and hardened as they are forced to the surface. The immense number of these cells surpasses comprehension. In one square inch of the cuticle, counting only those in a single layer, there are over a billion horny scales, each complete in itself.—*Harting.*

Value of the Cuticle.—In the palm of the hand, the sole of the foot, and other parts especially liable to injury, the cuticle is very thick. This is a most admirable provision for their protection.† By use, it becomes callous and horny. The boy who goes out barefoot for the first time, "treading as if on eggs,"

---

* We see how rapidly this change goes on by noticing how soon a stain of any kind disappears from the skin. A snake throws off its cuticle entire, and at regular intervals.

† We can hold the hand in strong brine with impunity, but the smart will quickly tell us when there is even a scratch in the skin. In vaccination, the matter must be inserted beneath the cuticle to take effect. Doubtless this membrane prevents many poisonous substances from entering the system.

can soon run where he pleases among thistles and over stones. The blacksmith handles hot iron without pain, while the mason lays stones and works in lime, without scratching or corroding his flesh.

The Complexion.—In the freshly-made cells on the lower side of the cuticle, is a pigment composed of tiny grains.* In the varying tint of this coloring-matter, lies the difference of hue between the blonde and the brunette, the European and the African. In the purest complexion, there is some of this pigment, which, however, disappears as the fresh, round, soft cells next the cutis change into the old, flat, horny scales at the surface.

Scars are white, because this part of the cuticle is not restored. The sun has a powerful effect upon the coloring-matter, and so we readily "tan" on exposure to its rays. If the color gathers in spots, it forms freckles.†

2. HAIR AND NAILS.

The Hair and the Nails are modified forms of the cuticle.

---

* These grains are about $\frac{1}{7000}$ of an inch in diameter, and, curiously enough, do not appear opaque but transparent and nearly colorless.—*Marshall.*

† This action of the sun on the pigment of the skin is very marked. Even among the Africans, the skin is observed to lose its intense black color in those who live for many months in the shades of the forest. It is said that Asiatic and African women confined within the walls of the harem, and thus secluded from the sun, are as fair as Europeans. Among the Jews who have settled in Northern Europe, are many of light complexion, while those who live in India are as dark as the Hindoos. The black pigment has been known to disappear during severe illness, and a lighter color to be developed in its place. Among the negroes, are sometimes found people who have no complexion, *i. e.*, there is no coloring-matter in their skin, hair, or the iris of their eyes. These persons are called Albinos.

Fig. 23.

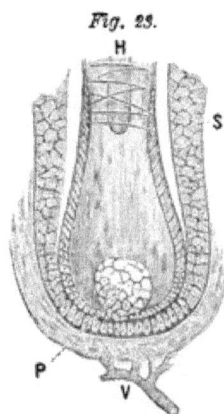

A hair magnified 600 diameters. S, the sac (follicle); P, the papilla, showing the cells and the bloodvessels (V).

The Hair is a protection from heat and cold, and shields the head from blows. It is found on nearly all parts of the body, except the palms of the hands and the soles of the feet. The outside of a hair is hard and compact, and consists of a layer of colorless scales, which overlie one another like the shingles of a house; the interior is porous,* and probably conveys the liquids by which it is nourished.

Each hair grows from a tiny bulb (papilla), which is an elevation of the cutis at the bottom of a little hollow in the skin. From the surface of this bulb, the hair is produced, like the cuticle, by the constant formation of new cells at the bottom. When the hair is pulled out, this bulb, if uninjured, will produce a new one; but, when once destroyed, it will never grow again.† The hair has been known to whiten in a single night by fear, fright, or nervous excitement. When the color has once changed, it cannot be restored.‡

* In order to examine a hair, it should be put on the slide of the microscope, and covered with a thin glass, while a few drops of alcohol are allowed to flow between the cover and the slide. This causes the air, which fills the hair and prevents our seeing its structure, to escape.

† The hair is said to grow after death. This is due to the fact that by the shrinking of the skin the part below the surface is caused to project, which is especially noticeable in the beard.

‡ Hair dyes or so-called "hair restorers" are almost invariably deleterious substances, depending for their coloring properties upon the action of lead or lunar caustic. Frequent instances of hair-poisoning have occurred, owing to the common use of such dangerous articles. If the growth of the hair be impaired, the general constitution or the skin needs treatment. This is the work of a skillful

Wherever hair exists, tiny muscles are found, interlaced among the fibers of the skin. These, when
contracting under the influ-
ence of cold or electricity,
pucker up the skin, and cause
the hair to stand on end.*
The hairs themselves are destitute of feeling. Nerves,
however, are found in the
hollows in which the hair is
rooted, and so one feels pain
when it is pulled. † Thus
the insensible hairs become
wonderfully delicate instruments to convey an impression of even the slightest
touch.

Next to the teeth and
bones, the hair is the least
destructible part of the body,
and its color is often preserved for many years after
the other portions have gone
to decay.‡

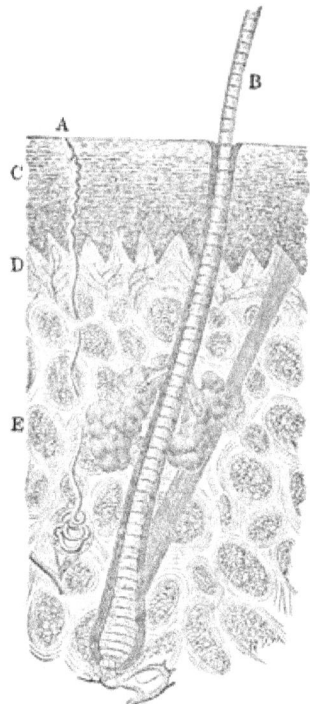

Fig. 24.

A, *a perspiratory tube with its
gland;* B, *a hair with a muscle
and two oil-glands;* C, *cuticle;*
D, *the papillæ;* and E, *fat-cells.*

physician, and not of a patent remedy. Dame Fashion has her repentant freaks as
well as her ruinous follies, and it is a healthful sign that the era of universal hair-
dyeing has been blotted out from her present calendar, and the gray hairs of age are
now honored with the highest place in "style" as well as in good sense and clean-
liness.

* In horses and other animals which are able to shake the whole skin, to drive
away the flies, this muscular tissue is much more fully developed than in man.

† These nerves are especially abundant in the whiskers of the cat, which are used
as feelers.

‡ Fine downy hairs so general upon the body have been detected in the little
fragments of skin found beneath the heads of the nails by which, centuries ago, certain robbers were fastened to the church doors, as a punishment for their sacrilege.

The Nails protect the ends of the tender finger, and toe, and give us power more firmly to grasp and easily to pick up any object we may desire. They enable us to perform a hundred little, mechanical acts which else were impossible. At the same time, their delicate color and beautiful outline give a finish of ornament to that exquisite instrument, the hand. The nail is firmly set in a groove (matrix) in the cuticle, from which it grows at the root in length * and from beneath in thickness. So long as the matrix at the root is uninjured, the nail will be replaced after any accident.

3. THE MUCOUS MEMBRANE.

Structure.—At the edges of the openings into the body, the skin seems to stop and give place to a tissue which is redder, more sensitive, more liable to bleed, and is moistened by a fluid, or mucus as it is called. Really, however, the skin does not cease, but passes into a more delicate covering of the same general composition, viz., an outer, hard, bloodless, insensible layer, and an inner, soft, sanguine, nervous one.† Thus every part of the body is contained in a kind of double bag, made of the tough skin on the outside, and the tender mucous membrane on the inside.

Connective Tissue.—The cutis and the correspond-

---

* By making a little mark on the nail near the root we can see week by week how rapidly this process goes on. and so form some idea of what a multitude of cells must be transformed into the horny matter of the nail.

† With a dull knife, we can scrape from the mucous membrane which lines the mouth some of the cuticle for examination under the microscope. In a similar way, we can obtain cuticle from the surface of the body for study and comparison.

ing layer of the mucous membrane consist chiefly of a fibrous substance interlaced like felt. It is called connective tissue, because it connects all the different parts of the body. It spreads from the cutis, invests muscles, bones, and cartilages, and thence passes into the mucous membrane. So thoroughly does it permeate the body, that, if the other tissues were destroyed, it would give a perfect model of every organ.* It can be seen in a piece of meat as a delicate substance lying between the layers of muscle, where it serves to bind together the numerous fibers of which they are composed.†

Connective tissue yields gelatine on boiling, and is the part which tans when hides are manufactured into leather. It is very elastic, so that when you remove your finger after pressing upon the skin, no indentation is left. ‡ It varies greatly in character, —from the mucous membrane, where it is soft and tender, to the ligaments and tendons which it largely composes, where it is strong and dense.§

---

* It is curious to notice how our body is wrapped in membrane. On the outside, is the skin protecting from exterior injury, and, on the inside, is the mucous membrane reaching from the lips to the innermost air-cell of the lungs. Every organ is enveloped in its membrane. Every bone has its sheath. Every socket is lined. Even the separate fibers of muscles have their covering tissue. The brain and the spinal cord are triply wrapped, while the eye is only a membranous globe filled with fluid. These membranes protect and support the organs they enfold, but, with that wise economy so characteristic of nature everywhere, they have also an important function to perform. They are the *filters* of the body. Through their pores pass alike the elements of growth, and the returning products of waste. On one side, bathed by the blood, they choose from it suitable food for the organ they envelop, and many of them in their tiny cells, by some mysterious process, form new products,—put the finishing touches, as it were, upon the material ere it is deposited in the body.

† Sometimes butchers blow air into veal, which fills the tiny cells of this tissue, and causes the meat to appear plump.

‡ In dropsy, this elasticity is lost by distension, and there is a kind of "pitting," as it is called, produced by pressure.

§ The leather made from this tissue varies as greatly, from the tough, thick ox-hide, to the soft, pliable kid and chamois skin.

Fat is deposited as an oil in the cells* of this tissue, just beneath the skin (Fig. 24), giving roundness and plumpness to the body, and acting as a powerful non-conductor for the retention of heat.† It collects as pads in the hollows of the bones, around the joints, and between the muscles, causing them to glide more easily upon each other. As marrow, it nourishes the skeleton, and also distributes the shock of any jar the limb may sustain.

It is noticeable, however, that fat does not gather within the cranium, the lungs, or the eyelids, where its accumulation would clog the organs.

4. THE TEETH.

The Teeth‡ are thirty-two in all,—there being eight in each half-jaw, similarly shaped and arranged. In each set of eight, the two nearest the middle of the jaw have wide, sharp, chisel-like edges, fit for cutting, and hence are called *incisors*. The next corresponds to the great tearing or holding tooth of the dog, and is styled the *canine*, or eye-tooth. The next two have broader crowns, with two

* So tiny are these cells, that there are over 63,000,000 in a cubic inch of fat.— *Valentine.*

† As the cells of the tissue are kept moistened, the liquid does not ooze out, but, on drying, comes to the surface. For this reason, a piece of fat feels oily when exposed to the air.

‡ Although the teeth are always found in connection with the skeleton, and are, therefore, figured as a part of it (Fig. 1), yet they do not properly belong to the bones of the body, and are merely set in the solid jaw to insure solidity. They are hard, and resemble bony matter, yet they are neither true bone nor are they formed in the same manner. "They are properly appendages of the mucous membrane, and are developed from it."—*Leidy.* "They belong to the Tegumentary System, which, speaking generally of animals, includes teeth, nails, horns, scales, and hairs."—*Marshall.* They are therefore classed with the mucous membrane, as are the nails and hair with the skin.

points, or cusps, and are hence termed the *bicuspids*. The remaining three are much broader, and, as they are used to crush the food, are called the *grinders*, or *molars*. The incisors and eye-teeth have one fang, or root, the others have two or three each.

The Milk-teeth.—We are provided with two sets of teeth. The first, or milk-teeth, are small and only twenty in number. In each half-jaw there are two incisors, one canine, and two molars. The middle incisors are usually cut about the age of seven months, the others at nine months, the first molars at twelve months, the canines at eighteen months, and the remaining molars at two or three years of age. The lower teeth precede the corresponding upper ones. The time often varies, but the order seldom.

Fig. 25.

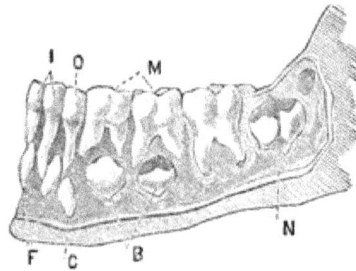

The teeth at the age of six and one-half years. I. the incisors ; O, the canine ; M, the molars ; the last molar is the first of the permanent teeth ; F, sacs of the permanent incisors ; C, of the canine ; B, of the bicuspids ; N, of the 2d molar ; the sac of the 3d molar is empty.—MARSHALL.

The Permanent Teeth.—At six years, when the first set is usually still perfect, the jaws contain the crowns of all the second, except the wisdom-teeth. About this age, to meet the wants of the growing body, the crowns of the permanent set begin to press against the roots of the milk-teeth, which, becoming absorbed, leave the loosened teeth to drop out, while the new ones rise and occupy their places.*

The central incisors appear at about seven years of age; the others at eight; the first bicuspids at nine, the second at ten; the canines at eleven or twelve; the second † molars at twelve or thirteen, and the last, or wisdom-teeth, are sometimes delayed until the twenty-second year, or even later.

Structure of the Teeth.—The interior of the tooth consists principally of *dentine*, a dense substance resembling bone.‡ The crown of the tooth, which is exposed to wear, is protected by a sheath of *enamel*. This is a hard, glistening, white substance, containing only two and a half per cent. of animal matter. The fang is covered by a thin layer of true bone (cement).

Fig. 27.

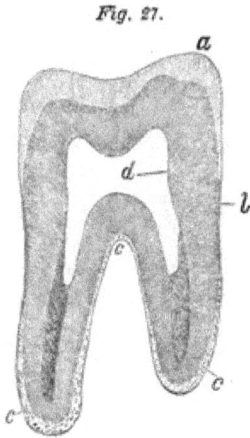

Vertical section of a Molar Tooth, moderately magnified. a, enamel of the crown, the lines of which indicate the arrangement of its columns; b, dentine; c, cement; d, pulp cavity.

At the center of the tooth is a cavity filled with a soft, reddish-white, pulpy substance full of blood-vessels and nerves. This pulp is very sensitive, and toothache is caused by its irritation.

* If the milk-teeth do not promptly loosen on the appearance of the second set, the former should be at once removed to permit the permanent teeth to assume their natural places. If any fail to come in regularly, or if they crowd the others, a competent dentist should be consulted.

† The first molar appears much earlier. (See Fig. 25.)

‡ In the tusk of the elephant this is known as ivory.

The Fitting of the Tooth into the Jaw is a most admirable contrivance. It is not set, like a nail in wood, having the fang in contact with the bone; but the socket is lined with a membrane which forms a soft cushion. While this is in a healthy state, it deadens the force of any shock, but, when inflamed, becomes the seat of excruciating pain.

The Decay of the Teeth * is commonly caused (1) by portions of the food which become entangled between them, and, on account of the heat and moisture, quickly decompose; and (2) by the saliva, as it evaporates, leaving on the teeth a sediment, which we call *tartar*. This collects organic matter that rapidly changes, and also affords a soil in which a sort of fungus speedily springs up. From both these causes, the breath becomes offensive, and the teeth are injured.

Preservation of the Teeth.—Children should early be taught to brush their teeth at least every morning with tepid water, and twice a week with soap and powdered orris-root. They should also be instructed to remove the particles of food from between the teeth, after each meal, by means of a a quill or wooden tooth-pick.

The enamel once injured is never restored, and the whole interior of the tooth is exposed to decay.

---

* Unlike the other portions of the body, there is no provision made for any change in the permanent teeth. That part, however, which is thus during life most liable to change, after death resists it the longest. In deep-sea dredgings teeth are found when all traces of the frame to which they belonged have disappeared. Yet hard and incorruptible as they seem, their permanence is only relative. Exposed to injury and disease, they break or decay. Even if they escape accident, they yet wear at the crown, are absorbed at the fang, and, in time, drop out, thus affording another of the many signs of the limitations Providence has fixed to the endurance of our bodies and the length of our lives.

We should not, therefore, crack hard nuts, bite thread, or use metal tooth-picks, gritty tooth-powders, or any acid which "sets the teeth on edge," i. e., that acts upon the enamel. It is well also to have the teeth examined yearly by a dentist, that any small orifice may be filled, and further decay prevented.

5. THE GLANDS OF THE SKIN.

I. The Oil Glands are clusters of tiny sacs which secrete an oil that flows along the duct to the root of the hair, and thence oozes out on the cuticle (Fig. 24.)* This is nature's efficient hair-dressing, and also keeps the skin soft and flexible. These glands are not usually found where there is no hair, as on the palm of the hand, and hence at those points only can water readily soak through the skin into the body. They are of considerable size on the face, especially about the nose. When obstructed, their contents become hard and dark-colored, and are vulgarly called " worms."†

II. The Perspiratory Glands are fine tubes about $\frac{1}{300}$ of an inch in diameter, and a quarter of an inch in length, which run through the cutis, and then coil up in little balls (Fig. 24). They are found in all parts of the body, and in almost incredible numbers. In the palm of the hand, there are about 2,800 in a single square inch. On the back of the neck

---

* This secretion is said to vary in different persons, and on that account the dog is enabled to trace his master by the scent.

† Though they are not alive, yet, under the microscope, they are sometimes found to contain a curious parasite called the pimple-mite, which is supposed to consume the superabundant secretion.

and trunk, where they are fewest, there are yet 400 to the square inch. The total number on the body of an adult is estimated at about two and a half million. If they were laid end to end, they would extend nearly ten miles.* The mouths of these glands—"pores," as we commonly call them—may be seen with a pocket lens along the fine ridges which cover the palm of the hand.

The Perspiration.—From these openings, there constantly passes a vapor, forming what we call the insensible perspiration. Exercise or heat causes it to flow more freely, when it condenses on the surface in drops. The perspiration consists of about ninety-nine parts water, and one part solid matter. The amount varies greatly, but on the average is, for an adult, not far from two pounds per day. The importance of this constant drainage has been shown by frequent experiments. Small animals, as the rabbit, when coated with varnish, die within twelve hours.†

The Absorbing Power of the Skin.—We have already described two uses of the skin : (1) Its *protective*, (2) its *exhaling*, and now we come to (3) its *absorbing* power. This is not so noticeable as the others, and yet it can be illustrated. Persons fre-

---

* The current statement, that they would extend twenty-eight miles, is undoubtedly an exaggeration. Krause estimates the total number at 2,381,248, and the length of each coil, when unraveled, at $\frac{1}{27}$ of an inch, which would make the total length much less than even the statement in the text. Seguin states that the proportion of impurities thrown off by the skin and the lungs, is eleven to seven.

† On an occasion of great solemnity, Pope Leo X. caused a child to be completely covered with gold leaf, closely applied to the skin, so as to represent, according to the idea of the age, the golden glory of an angel or seraph. Within a few hours after this pageant the child died. The ignorant common people of those days attributed the death to the anger of the Deity, and looked upon it as an evil omen.

quently poison their hands with the common wood-
ivy. Contagious diseases are caught by touching a
patient, or even his clothing, especially if there be a
crack in the cuticle.* Painters absorb so much lead
through the pores of their hands that they are
attacked with colic.† Snuff and lard are frequently
rubbed on the chest of a child suffering with the
croup, to produce vomiting. It is said that seamen
in want of water drench their clothing in salt spray,
and the skin will absorb enough moisture to quench
thirst (see Lymphatic System).

By carefully conducted experiments, it has been
found that the skin acts in the same way as the
lungs (see Respiration) in absorbing oxygen from
the air, and giving off carbonic acid to a small
but appreciable amount. Indeed, the skin has not
inaptly been styled the third lung.‡

6. HYGIENE.

Hints about Washing and Bathing.—The moment
of rising from bed is the proper time for the full
wash or bath with which one should commence the
day. The body is then warm, and can endure mod-

* If one is called upon to handle a dead body, it is well, especially if the person
has died of a contagious disease, to rub the hand with lard or olive-oil. Poisonous
matter has been fatally absorbed through the breaking of the cuticle by a hang-nail,
or a simple scratch. There is a story that Buonaparte, when a lieutenant of artillery,
in the heat of battle, seized the rammer and worked the gun of an artillery-man who
had fallen. From the wood which the soldier had used, Buonaparte absorbed a poison
that gave him a skin-disease, by which he was annoyed the remainder of his life.

† Cosmetics, hair-dyes, etc., are exceedingly injurious, not only because they tend
to fill the pores of the skin, but because they often contain poisonous matters that
may be absorbed into the system, especially if they are in a solution.

‡ In some of the lower animals, it plays a still more important part. Frogs,
deprived of their lungs, breathe with almost undiminished activity, and often sur-
vive for days.

erately cold water better than at any other time ; it is relaxed, and needs bracing; and the nerves, deadened by the night's repose, require a gentle stimulus. If the system be strong enough to resist the shock, cold water is the most invigorating ; if not, a tepid bath will answer.*

Before dressing, the whole body should be thoroughly rubbed with a coarse towel or flesh-brush. At first, the friction may be unpleasant, but this sensitiveness will soon be overcome, and the keenest pleasure be felt in the lively glow which follows. A bath should not be taken just before nor immediately after a meal, as it will interfere with the digestion of the food. Soap should be employed occasionally, but its frequent use tends to make the skin dry and hard.

Reaction.—After taking a cold bath, there should be a prompt reaction. When the surface is chilled by cold water, the blood sets to the heart and other vital organs, exciting them to more vigorous action, and then, being thrown back to the surface, it reddens, warms, and stimulates the skin to an unwonted degree. This is called the reaction, and in

---

* Many persons have not the conveniences for a bath. To them, the following plan, which the author has daily employed for years, is commended. The necessities are: a basin full of soft water, a mild soap, a large sponge or a piece of flannel, and two towels—one soft, the other rough. The temperature of the water should vary with the season of the year—cold in summer and tepid in winter. Rub quickly the entire body with the wet sponge or flannel. (If more agreeable, wash and wipe only a part at a time, protecting the rest in cold weather with portions of clothing.) Dry the skin gently with a soft towel, and when quite dry, with the rough towel or flesh-brush rub the body briskly four or five minutes till the skin is all aglow. The chest and abdomen need the principal rubbing. The roughness of the towel should be accommodated to the condition of the skin. Enough friction, however, must be given to produce at least a gentle warmth, indicative of the reaction necessary to prevent subsequent chill or languor. An invalid will find it exceedingly beneficial if a stout, vigorous person produce the reaction by rubbing with the hands.

it lies the invigorating influence of the cold bath. If, on the contrary, the skin be heated by a hot bath, the blood is drawn to the surface, less blood goes to the heart, the circulation decreases, and languor ensues. A dash of cold water is both necessary and refreshing at its close.*

If, after a cold bath, there be felt no glow of warmth, but only a chilliness and depression, we are thereby warned that either proper means were not taken to bring on this reaction, or that the circulation is not vigorous enough to make such a bath beneficial. The general effect of a cool bath is exhilarating, and that of a warm one depressing.† Hence the latter should not ordinarily be taken oftener than once a week, while the former may be enjoyed daily.

Sea-bathing is exceedingly stimulating, on account of the action of the salt and the exciting surroundings. Twenty minutes is the utmost limit for bathing or swimming in salt or fresh water. A chilly sensation should be the signal for instant removal. It is better to leave while the glow and buoyancy which follow the first plunge are still felt. Gentle exercise after a bath is beneficial.

---

* The Russians are very fond of vapor baths, taken in the following manner. A large room is heated by stoves. Red-hot stones being brought in, water is thrown upon them, filling the room with steam. The bathers sit on benches until they perspire profusely, when they are rubbed with soapsuds and dashed with cold water. Sometimes, while in this state of excessive perspiration, they run out of doors and leap into snow-banks.

† The sudden plunge into a cold bath is good for the strong and healthy, but too severe for the delicate. One should always wet first the face, neck, and chest. It is extremely injurious to stand in a bath with only the feet and the lower limbs covered by the water, for the blood is thus sent from the extremities to the heart and internal organs, and they become so burdened that reaction may be out of their power. A brisk walk, or a thorough rubbing of the skin, before a cold bath or swim, adds greatly to its value and pleasure.

Clothing in winter, to keep us warm, should repel the external cold and retain the heat of the body. In summer, to keep us cool, it should not absorb the rays of the sun, and should permit the passage of the heat of the body. At all seasons, it should be porous, to give ready escape to the perspiration, and a free admission of air to the skin. We can readily apply these essential conditions to the different kinds of clothing.

Linen is soft to the touch, and is a good conductor of heat. Hence it is pleasant for summer wear, but, being apt to chill the surface too rapidly, it should not be worn next the skin.

Cotton is a poorer conductor of heat and absorber of moisture, and is therefore warmer than linen. It is sufficiently cool for summer wear, and affords better protection against sudden changes.

Woolen absorbs moisture slowly, and contains much air in its pores. It is therefore a poor conductor of heat, and guards the wearer against the vicissitudes of our climate.

The outer clothing may be adapted largely to ornament, and may be varied to suit our fancy and the requirements of society. But the body should be protected by plentiful under-clothing, which should be of itself sufficient to keep us warm. Flannel should be worn next the skin at all times, except in the heat of summer, when cotton flannel may be substituted. In the coldest weather, it should be doubled. Its roughness is sometimes disagreeable, but habit soon overcomes this sensitiveness, and renders it exceedingly grateful.

Light-colored clothing is not only cooler in summer, but warmer in winter. As the warmth of clothing depends greatly on the amount of air contained in its fibers, fine, loose, porous cloth with a plenty of nap is best for winter wear. Firm and heavy goods are not necessarily the warmest. Furs are the perfection of winter clothing, since they combine warmth with lightness. Two light woolen garments are warmer than one heavy one, as there is between them a layer of non-conducting air.

All the body except the head should be equally protected by clothing. Whatever fashion may dictate, no part covered to-day can be uncovered to-night or to-morrow, except at the peril of health. It is a most barbarous and cruel custom to leave the limbs of little children unprotected, when adults would shiver at the very thought of exposure. Equally so is it for children to be thinly clad for the purpose of hardening them. To go shivering with cold is not the way to increase one's power of endurance. The system is made more vigorous by exercise and food ; not by exposure. In winter, there is more fear of too little than too much clothing. Above all, the feet need heavy shoes with thick soles, and rubbers when it is damp. At night, and after exercise, we require extra clothing.

Diseases, etc.—ERYSIPELAS is an inflammation (see Inflammation) of the skin, and often begins in a spot not larger than a pin-head. which spreads with great rapidity. It is very commonly checked by the application of a solution of iodine. The burning and contracting sensation may be relieved by cloths wrung out of hot water.

2. DROPSY is a disease in which there is an accumulation of water in the system. On account of the free passage between the cells of the connective tissue, this liquid gradually settles into the feet when the person is standing ; on reclining, the equilibrium is restored.

3. CORNS are a thickened part of the cuticle, caused by pressure or friction. They most frequently occur on the feet ; but are produced on the shoemaker's knee by constant hammering, and on the soldier's shoulder by the rubbing of his musket. This hard portion irritates the sensitive cutis beneath, and so causes pain. By soaking the feet in hot water, the corn will be softened, when it may be pared with a sharp knife. If the cause be removed, the corn will not return.

4. IN-GROWING NAILS are caused by pressure, which forces the edge of the toe-nail into the flesh. They may be cured by carefully cutting away the part which has mal-grown, and then scraping the back of the nail till it is thin and no longer resists the pressure. The two portions, uniting, will draw away the nail from the flesh at the edge. They are prevented by paring the nail straight across, thus making the corners right angles, and by wearing broad shoes.

5. WARTS are overgrown papillæ (Fig. 24.). They may generally be removed by the application of glacial acetic acid, or a drop of nitric acid, repeated until the entire structure is softened. Care must be

taken to keep the acid from touching the neighboring skin. The capricious character of warts has given rise to the popular delusion concerning the influence of charms upon them.

6. CHILBLAIN is a local inflammation affecting generally the feet, the hands, or the lobes of the ear. Liability to it usually passes away with manhood. It is not caused by "freezing the feet," as many suppose, though attacks are brought on, or aggravated, by exposure to cold and by sudden warming. It is subject to daily congestion (see Congestion), manifested by itching, soreness, etc., commonly occurring at night. The best preventive is a uniform temperature, and careful protection against the cold by warm, loose, and plentiful clothing, especially for the feet.

7. WENS are caused by an unnatural activity of the arteries, which deposit more nutriment than is needed. Physicians " scatter them," as it is termed, by stimulating the absorbents to carry off the excess. A boil often disappears without "coming to a head" in a somewhat similar way, i. e., by the renewed activity of the absorbing vessels.

PRACTICAL QUESTIONS.

1. If a hair be plucked out, will it grow again ?

2. What causes the hair to "stand on end" when we are frightened ?

3. Why is the skin roughened by riding in the cold ?

4. Why is the back of a washer-woman's hand less water-soaked than the palm?

5. What would be the length of the perspiratory tubes in a single square inch of the palm, if placed end to end?

6. What colored clothing is best adapted to all seasons?

7. What is the effect of paint and powder on the skin?

8. Is water-proof clothing healthful for constant wear?

9. Why are rubbers cold to the feet?

10. Why does the heat seem oppressive when the air is moist?

11. Why is friction of the skin invigorating after a cold bath?

12. Why does the hair of domestic animals become roughened in winter?

13. Why do fowls shake out their feathers erect before they perch for the night?

14. How can an extensive burn cause death by congestion of the lungs?

15. Why do we perspire so profusely after drinking cold water?

16. What are the best means of preventing skin-diseases, colds, and rheumatism?

17. What causes the difference between the hard hand of a black-smith and the soft hand of a woman?

18. Why should a painter avoid getting paint on the palm of his hand?

19. Why should we not use the soap or the soiled towel at a hotel?

20. Which teeth cut like a pair of scissors?

21. Which teeth cut like a chisel?

22. Which should be clothed the warmer, a merchant or a farmer?

23. Why should we not crack nuts with our teeth?

24. Do the edges of the upper and the lower teeth meet?

25. When fatigued, would you take a cold bath?

26. Why is the outer surface of a kid glove finer than the inner?

27. Why will a brunette endure the sun's rays better than a blonde?

28. Does patent-leather form a healthful covering for the feet?

29. Why are men more frequently bald than women?

30. On what part of the head does baldness commonly occur? Why?

31. What does the combination in our teeth of canines and grinders suggest as to the character of our food?

32. Is a staid, formal promenade suitable exercise?

33. Is there any danger in changing the warm clothing of our daily wear for the thin one of a party?

34. Should we retain our overcoat, shawl, or furs when we come into a warm room?

35. Which should bathe the oftener, students or out-door laborers?

36. Is abundant perspiration injurious?

37. How often should the ablution of the entire body be performed?

38. Why is cold water better than warm, for our daily ablution?

39. Why should our clothing always fit loosely?

40. Why should we take special pains to avoid clothing that is colored by poisonous dye-stuffs?

41. What general principles should guide us as to the length and frequency of baths in salt or fresh water?

42. What is the beneficial effect of exercise upon the functions of the skin?

43. How can we best show our admiration and respect for the human body?

44. Why is the scar of a severe wound upon a negro sometimes white?

IV.

RESPIRATION

AND

THE VOICE.

" The smooth soft air with pulse-like waves
Flows murmuring through its hidden caves,
Whose streams of brightening purple rush,
Fired with a new and livelier blush ;
While all their burthen of decay
The ebbing current steals away."

BLACKBOARD ANALYSIS.

RESPIRATION AND THE VOICE.

1. ORGANS OF VOICE
1. The Larynx.
2. The Vocal Cords.
3. Different Tones of Voice.
4. Speech.
5. Formation of Vocal Sounds.

2. ORGANS OF RESPIRATION.
1. The Trachea.
2. The Bronchial Tubes.
3. The Cells.
4. The Lung-wrapping.
5. The Cilia.

3. HOW WE BREATHE
1. Inspiration.
2. Expiration.

4. MODIFICATIONS OF THE BREATH.
1. Sighing.
2. Coughing.
3. Sneezing.
4. Snoring.
5. Laughing, and Crying.
6. Hiccough.
7. Yawning.

5. CAPACITY OF THE LUNGS.

6. HYGIENE.
1. The Need of Air.
2. Action of Air in the Lungs.
3. Tests of the Breath.
4. Analysis of expired Air.
5. Effect of re-breathed Air.
6. Concerning the Need of Ventilation.
   a. The Sources of Impurity.
   b. The Sick-room.
   c. The Sitting-room.
   d. The Bed-room.
   e. The Church.
   f. The School-room.
   g. How we should ventilate.

7. THE WONDERS OF RESPIRATION.

8. DISEASES
1. Constriction of the Lungs.
2. Bronchitis.
3. Pleurisy.
4. Pneumonia.
5. Consumption.
6. Asphyxia.
7. Diphtheria.
8. Croup.
9. Stammering.

RESPIRATION

AND

THE VOICE.

THE ORGANS of Respiration and the Voice are the *larynx*, the *trachea*, and the *lungs*.

Description of the Organs of the Voice.—1. THE LARYNX.—In the neck, is a prominence sometimes called Adam's apple. It is the front of the *larynx*. This is a small triangular, cartilaginous box, placed just behind the tongue, and at the top of the windpipe. The opening into it from the throat is called the *glottis ;* and the cover, the *epiglottis* (*epi*, upon ; *glotta*, the tongue). The latter is a spoon-shaped lid, which opens when we breathe, but, by a nice arrangement, shuts when we try to swallow, and so lets our food slip over it into the *œsophagus* (e-sof'-a-gus), the tube leading from the pharynx to the stomach (Fig. 27).

If we laugh or talk when we swallow, our food is apt to "go the wrong way," *i. e.*, little particles pass into the larynx, and the tickling sensation which

Fig. 27.

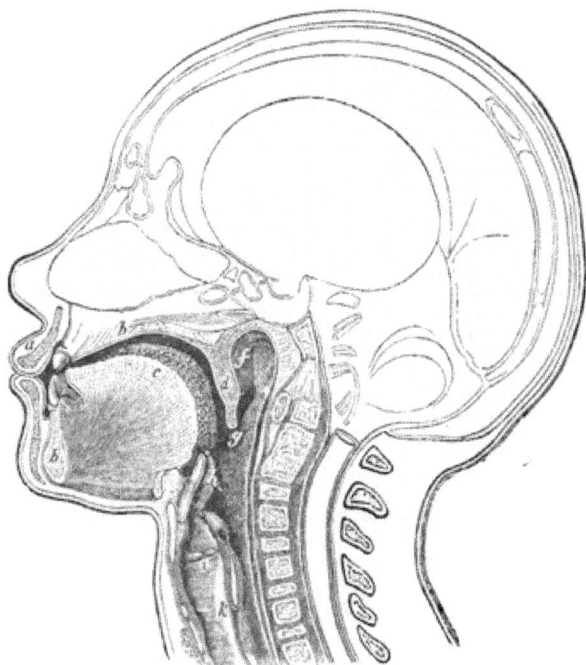

Passage to the Œsophagus and Windpipe: c, the tongue; d, the soft palate. ending in g, the uvula; h, the epiglottis; i, the glottis; I, the œsophagus; f, the pharynx.

they produce forces us to cough in order to expel the intruders.

2. THE VOCAL CORDS.—On each side of the *glottis* are the so-called *vocal cords*. They are not really cords, but merely elastic membranes projecting from the sides of the box across the opening.* When not in use, they spread apart and leave a V-shaped orifice (Fig. 28), through which the air passes to and from the lungs.   If the cords are tightened, the edges

---

* The cartilages and vocal cords of the larynx may be readily seen in that of the ox or sheep. If the flesh be cut off, the cartilages will dry, and will keep for years.

approach sometimes within $\frac{1}{100}$ of an inch of each other, and, being thrown into vibration, cause corresponding vibrations in the current of air. Thus sound is produced in the same manner as by the vibrations of the tongues of an accordeon, or the strings of a violin, only in this case the strings are scarcely an inch long.

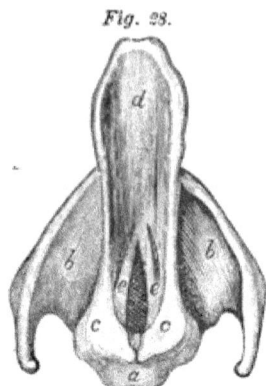

Fig. 28.

e, e, the vocal cords ; d, the epiglottis.

Different Tones of the Voice. —The higher tones of the voice are produced when the cords are short, tight, and closely in contact ; the lower, by the opposite conditions. Loudness is regulated by the quantity of air and force of expulsion. A falsetto voice is thought to be the result of a peculiarity in the pharynx (Fig. 27) at the back part of the nose ; it is more probably produced by some muscular manœuvre not yet fully understood. When boys are about fourteen years of age, the larynx enlarges, and the cords grow proportionately longer and coarser ; hence, the voice becomes deeper, or, as we say, "changes." The peculiar harshness of the voice at this time seems to be due to a congestion of the mucous membrane of the cords. The change may occur very suddenly, the voice breaking in a single night.

Speech is voice modulated by the lips, tongue,*

---

* The tongue is styled the "unruly member," and held responsible for all the tattling of the world ; but when the tongue is removed, the adjacent organs in some way largely supply the deficiency, so that speech is still possible. Huxley describes the conversation of a man who had two and one-half inches of his tongue preserved

palate, and teeth.* It is commonly associated with
the voice, but is not necessary to it; for when we
whisper we articulate the words, although there is
no vocalization, *i. e.*, no action of the larynx.†

Fig. 29.

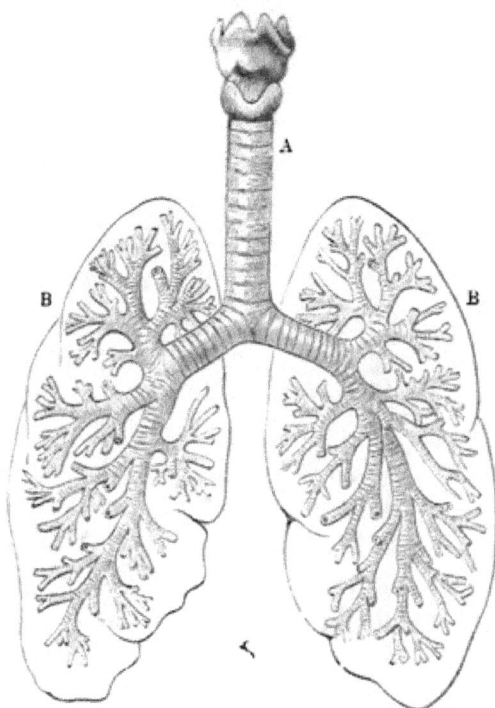

The Lungs, showing the Larynx. A, the windpipe ; B, the bronchial tubes.

in spirits, and yet could converse intelligibly. Only the two letters *t* and *d* were
beyond his power; hence, tin became " tin," and dog became "thog."

    * An artificial larynx may be made by using elastic bands to represent the vocal
cords, and by placing above them chambers which by their resonance will produce
the same effect as the cavities lying above the larynx. An artificial speaking-
machine was constructed by Kempelen, which could pronounce such sentences as,
"I love you with all my heart," in different languages, by simply touching the
proper keys.

    † We can observe this by placing the hand on the throat, and noticing the absence
of vibrations when we whisper, and their presence when we talk. The difference
between vocalization and non-vocalization is seen in a sigh and a groan, the latter

Formation of Vocal Sounds.—The method of modulating voice into speech may be seen by producing the pure vowel sounds *a*, *e*, etc., from one expiration, the mouth being kept open while the form of the aperture is changed for each vowel by the tongue and the lips. *H* is only an explosion, or forcible throwing of a vowel sound from the mouth.*

The consonants, or short sounds, may also be made without interrupting the current of air, by various modifications of the vocal organs. In sounding singly any one of the letters, we can detect its peculiar requirements. Thus *m* and *n* can be made only by blocking the air in the mouth and sending it through the nose ; *l* lets the air escape at the sides of the tongue ; *r* needs a vibratory movement of the tongue ; *b* and *p* stop the breath at the lips ; *d* and *t*, at the back of the palate. Consonants like *b* and *d* are abrupt, or, like *l* and *s*, continuous. Those made by the lips are termed *labials ;* those by pressing the tongue against the teeth, *dentals ;* those by the tongue, *linguals.*

The child gains speech slowly, first learning to pronounce the vowel *a*, the consonants *b*, *m*, and *p*, and then their unions—*ba, ma, pa.*

Description of the Organs of Respiration.—Beneath the larynx is the wind-pipe, or *trachea* (see Fig. 29), so called because of its roughness. It is strength-

---

being the former vocalized: Whistling is a pure mouth-sound. and does not depend on the voice. Laughter is vocal, being the aspirated vowels, a, e, or o, convulsively repeated.

* When, in sounding a vowel, the sound coincides with a sudden change in the position of the vocal cords from one of divergence to one of approximation, the vowel is pronounced with the *spiritus asper.* When the vocal cords are brought together before the blast of air begins, the vowel is pronounced with the *spiritus lenis.—Foster.*

ened by C-shaped cartilages with the openings
behind, where they are attached to the œsophagus.
At the lower end, the tra-

Fig. 30.

chea divides into two
branches, called the right
and left *bronchi*. These
subdivide in the small
bronchial tubes, which
ramify through the lungs
like the branches of a
tree, the tiny twigs of
which at last end in clusters of cells so small that
there are 600,000,000 in
all. This cellular struc-
ture renders the lungs

Bronchial Tubes, with clusters of cells.

exceedingly soft, elastic,
and sponge-like.*

The stiff, cartilaginous rings, so noticeable in the
rough surface of the trachea and the bronchi, disappear as we reach the smaller bronchial tubes, so
that while the former are kept constantly open for
the free admission of air, the latter are provided with
elastic fibers by which they may be almost closed.

Wrappings of the Lungs.—The lungs are invested
with a double covering—the *pleura*—one layer being
attached to the lungs and the other to the walls of
the chest. It secretes a fluid which lubricates it, so
that the layers glide upon each other with perfect

---

* The lungs of slaughtered animals are vulgarly called "lights," probably on
account of their lightness. They are similar in structure to those of man. They
will float on water, and if a small piece be forcibly squeezed between the fingers
(notice the creaking sound it gives), it will still retain sufficient air to make it
buoyant.

Fig. 31.

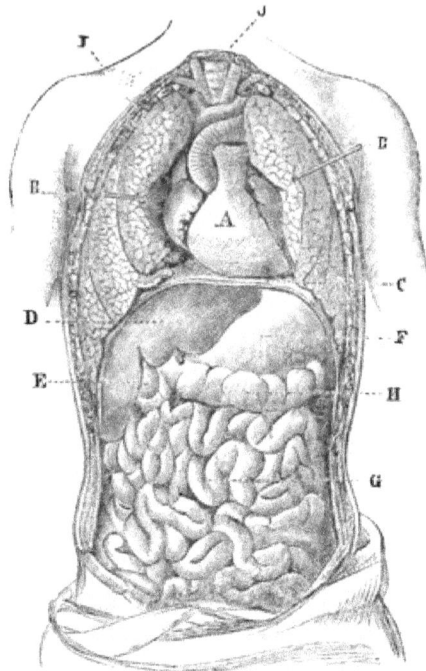

A, *the heart* ; B. *the lungs drawn aside to show the internal organs ;* C, *the diaphragm ;* D, *the liver ;* E, *the gall cyst ;* F, *the stomach ;* G, *the small intestine ;* H, *the transverse colon.*

ease.* The lungs are lined with mucous membrane, exceedingly delicate and sensitive to the presence of anything except pure air. We have all noticed this when we have breathed anything offensive.

The Cilia.—Along the air passages are minute filaments (*cilia*, Fig. 32), which are in constant motion, like a field of grain stirred by a gentle breeze. They serve to fan the air in the lungs, and produce an outward current, which is useful in catching dust and fine particles swept inward with the breath.

* These pleural sacs are distinct and closed ; hence, when the ribs are raised, a partial vacuum being formed in the sacs, air rushes in, and distends the pulmonary lobules.

Fig. 32.

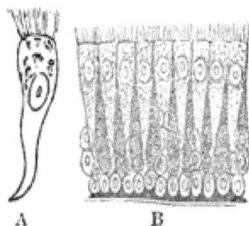

A B

B, a section of the mucous membrane, showing the cilia rising from the peculiar epithelial cells on the outside of the mucous membrane lining the tubes ; A, a single cell more highly magnified.

How we Breathe.—Respiration consists of two acts—taking in the air, or *inspiration*, and expelling the air, or *expiration*.

1. INSPIRATION. — When we draw in a full breath, we straighten the spine and throw the head and shoulders back, so as to give the greatest advantage to the muscles.* At the same time, the diaphragm † descends and presses the walls of the abdomen outward. Both these processes increase the size of the chest. Thereupon, the elastic lungs expand to occupy the extra space, while the air, rushing in through the windpipe, pours along the bronchial tubes and crowds into every cell.‡

2. EXPIRATION.—When we forcibly expel the air

* If we examine the bony cage of the thorax or chest in Fig. 8. we shall see that the position of the ribs may alter its capacity in two ways. 1. As they run obliquely downward from the spine, if the sternum or breast-bone be lifted in front, the diameter of the chest will be increased. 2. The ribs are fastened by elastic cartilages, which stretch as the muscles that lift the ribs contract, and so increase the breadth of the chest.

† The diaphragm is the muscular partition between the chest and the abdomen. It is always convex toward the former and concave toward the latter (Fig. 31). Long muscular fibers extend from its center toward the ribs in front and the spine at the back. When these contract, they depress and flatten the diaphragm ; when they relax, it becomes convex again. In the former case, the bowels are pressed downward and the abdomen pushed outward ; in the latter, the bowels spring upward, and the abdomen is drawn inward.

‡ It is said that in drawing a full breath, the muscles exert a force equal to raising a weight of 750 pounds. When we are about to make a great effort, as in striking a heavy blow, we naturally take a deep inspiration, and shut the glottis. The confined air makes the chest tense and firm, and enables us to exert a greater force. As we let slip the blow, the glottis opens and the air escapes, often with a curious aspirated sound, as is noticeable in workmen To make a good shot with a rifle, we should take aim with a full chest and tight breath, since then the arms will have a steadier support.

from our lungs, the operation is reversed. We bend forward, draw in the walls of the abdomen, and press the diaphragm upward, while the ribs are pulled downward,—all together diminishing the size of the chest, and forcing the air outward.

Ordinary, quiet breathing is performed mainly by the diaphragm,—one breath to every four beats of the heart, or eighteen per minute.

Modifications of the Breath.—*Sighing* is merely a prolonged inspiration followed by an audible expiration. *Coughing* is a violent expiration in which the air is driven through the mouth. *Sneezing* differs from coughing, the air being forced through the nose. *Snoring* is a sleeping accompaniment, in which the air passes through both nose and mouth. The peculiar sound is produced by the palate flapping in this divided current of air, and so throwing it into vibration. *Laughing* and *crying* are very much alike. The expression of the face is necessary to distinguish between them. The sounds are produced by short, rapid contractions of the diaphragm. *Hiccough* is confined to inspiration. It is caused by a contraction of the diaphragm and a constriction of the glottis ; the current of air just entering, as it strikes the closed glottis, gives rise to the well-known sound. *Yawning*, or *gaping*, is like sighing.* It is distinguished by a wide opening of the mouth and a deep, profound inspiration. Both processes furnish additional air, and therefore probably meet a demand

* Their usefulness lies in bringing up the arrears, as it were, of respiration, when it has fallen behindhand either through fatigue or close attention to other occupation. The stretching of the jaws and limbs may also serve to equalize the nervous influence, certain muscles having become uneasy on account of being stretched or contracted for a long time.

of the system for more oxygen. Frequently, however, they are like laughing, sobbing, etc., merely a sort of contagion, which runs through an audience, and seems almost irresistible.

The Capacity of the Lungs.—If we take a deep inspiration, and then forcibly exhale all the air we can expel from the lungs, this amount, which is termed the *breathing capacity,* will bear a very close correspondence to our stature. For a man of medium height (5 ft. 8 in.) it will be about 230 cubic inches,* or a gallon, and for each inch of height between five and six feet there will be an increase of eight cubic inches. In addition, it is found that the lungs contain about 100 cubic inches which cannot be expelled. thus making their entire contents about 330 cubic inches, or eleven pints. The extra amount always on hand in the lungs is of great value, since thereby the action of the air goes on continuously, even during a violent expiration. In ordinary breathing, only about twenty or thirty cubic inches (less than a pint) of air pass in and out.

The Need of Air.—The body needs food, clothing. sunshine, bathing, and drink ; but none of these wants is so pressing as that for air. The other demands may be met by occasional supplies, but air must be furnished every moment or we die. Now the vital element of the atmosphere is oxygen gas.†

---

* Of this amount, 100 cubic inches can be forced in only by an extra effort, and is available for emergencies, or for purposes of training, as in singing, climbing. etc. It is of great importance, since, if the capacity of the lungs only equaled our daily wants, the least obstruction would prove fatal.

† See *Steele's Chemistry.* page 43. The atmosphere consists of one-fifth oxygen and four-fifths nitrogen. The former is the active element ; and the latter, the passive. Oxygen alone would be too stimulating, and must be restrained by the neutral nitrogen.

This is a stimulating, life-giving principle. No tonic will so invigorate as a few full, deep breaths of cold, pure air. Every organ will glow with the energy of the fiery oxygen.

Action of the Air in the Lungs.—In the delicate cells of the lungs, the air gives up its oxygen to the blood, and receives in turn carbonic-acid gas and water, foul with waste matter which the blood has picked up in its circulation through the body. The blood, thus purified and laden with the inspiring oxygen, goes bounding through the system, while the air we exhale carries off the impurities. In this process, the blood changes from purple to red. If we examine our breath, we can readily see what it has removed from the blood.

Tests of the Breath.—1. Breathe into a jar, and on lowering into it a lighted candle, the flame will be instantly extinguished ; thus indicating the presence of carbonic-acid gas. 2. Breathe upon a mirror, and a film of moisture will show the vapor.* 3. If the breath be confined in a bottle for a time, the animal matter will decompose and give off an offensive odor.

Analysis of the Expired Air shows that it has lost about twenty-five per cent. of its oxygen, and gained an equal amount of carbonic-acid gas, besides moisture, and organic impurities. Our breath, then, is air robbed of its vitality, and containing in its

---

* There is a close relation between the functions of the skin, the lungs, and the kidneys—the scavengers of the body. They all carry off water from the blood, and when the function of one of the three is, in this respect, interfered with, the others are called upon to perform its functions. When the function of perspiration is deranged, the lungs and kidneys are required to perform heavier duty, and this may lead to disease (p. 61).

place a gas as fatal to life * as it is to a flame, and
effete matter which is disagreeable to the smell,
injurious to the health, and may contain the germs
of disease.

The Evil Effect of Re-breathing the air cannot be
over-estimated. We take back into our bodies that
which has just been rejected. The blood thereupon
leaves the lungs, bearing, not the invigorating
oxygen, but refuse matter to obstruct the whole system. We soon feel the effect. The muscles become
inactive. The blood stagnates. The heart acts
slowly. The food is undigested. The brain is
clogged. The head aches. Instances of fatal results
are only too frequent.† The constant breathing of
even the slightly-impure air of our houses cannot
but tend to undermine the health. The blood is not
purified, and is thus in a condition to receive the
seeds of disease at any time. The system uninspired
by the energizing oxygen is sensitive to cold. The
pale cheek, the lustreless eye, the languid step,
speak but too plainly of oxygen starvation. In such
a soil, catarrh, scrofula, and consumption run riot.‡

---

* Carbonic-acid gas cannot be breathed when undiluted, as the glottis closes and
forbids its passage into the lungs. Air containing only three or four per cent, acts
as a narcotic poison (*Miller*), and a much smaller proportion will have an injurious
effect. The great danger. however, lies in the organic particles constantly exhaled
from the lungs and the skin, which, it is believed. are often direct and active
poisons.

† During the English war in India in the last century, 146 prisoners were shut up
in a room scarcely large enough to hold them. The air could enter only by two narrow windows. At the end of eight hours, but twenty-three persons remained alive,
and these were in a most deplorable condition. This prison is well called "The
Black Hole of Calcutta."—Percy relates that after the battle of Austerlitz, 300 Russian prisoners were confined in a cavern. where 260 of them perished in a few hours.
—The stupid captain of the ship Londonderry, during a storm at sea, shut the
hatches. There were only seven cubic feet of space left for each person, and in six
hours ninety of the passengers were dead.

‡ "One not very strong, or unable powerfully to resist conditions unfavorable to

Concerning the Need of Ventilation.—The foul air which passes off from the lungs and through the pores of the skin does not fall to the floor, but diffuses itself through the surrounding atmosphere. A single breath will to a trifling but certain extent taint the air of a whole room.* A light will vitiate as much air as a dozen persons. Many breaths and lights therefore rapidly unfit the air for our use.

The perfection of ventilation is reached when the air of a room is as pure as that out of doors. To accomplish this result, it is necessary to allow for each person 600 cubic feet of space, while ventilation is still going on in the best manner known.

In spite of these well-known facts, scarcely any pains are taken to supply fresh air, while the doors and windows where the life-giving oxygen might creep in are hermetically stopped.

How often is this true of the sick-room. Yet here the danger of bad air is intensified. The expired breath of the patient is peculiarly threatening to himself as well as to others. Nature is seeking to throw off the poison of the disease. The scavengers of the body are all at work. The breath and the insensible perspiration are loaded with impurities.†

health, and with a predisposition to lung disease, will be sure, sooner or later, by partial lung-starvation and blood-poisoning, to develop pulmonary consumption. *The lack of what is so abundant and so cheap—good, pure air—is unquestionably the one great cause of this terrible disease."—Black's Ten Laws of Health.*

* This grows out of a well-known philosophical principle called the Diffusion of Gases, whereby two gases tend to mix in exact proportions, no matter what may be the quantity of each.—*Steele's Chemistry,* p. 96, and *Physics,* p. 49.

† The floating dust in the air, revealed to us by the sunbeam shining through a crack in the blinds, shows the abundance of these impurities, and also the presence of germs which, lodging in the lungs, may implant disease, unless thrown off by a vigorous constitution. "On uncovering a scarlet-fever patient, a cloud of fine dust is seen to rise from the body—contagious dust, that for days will retain its poisonous properties."—*Youmans.*

The odor is oftentimes exceedingly offensive. Sick and well alike need an abundance of fresh air. But, too often, it is the only want not supplied.

Our sitting-rooms, heated by furnaces or red-hot stoves, generally have no means of ventilation, or, if provided, they are seldom used. A window is occasionally dropped to give a little relief, as if pure air were a rarity, and must be doled out to the suffering lungs in morsels, instead of full and constant draughts. The inmates are starved by scanty lung-food, and stupefied by foul air. The process goes on year by year. The weakened and poisoned body at last succumbs to disease, while we, in our blindness and ignorance, talk of the mysterious Providence which thus untimely cuts down the brightest intellects. The truth is, death is often simply the penalty for violating nature's laws. Bad air begets disease ; disease begets death.

In our churches, the foul air left by the congregation on Sunday is shut up during the week, and heated for the next Lord's day, when the people assemble to re-breathe the polluted atmosphere. They are thus forced, with every breath they take, to violate the physical laws of Him whom they meet to worship,—laws written not 3000 years ago upon Mount Sinai on tables of stone, but to-day engraved in the constitution of their own living, breathing bodies. On brains benumbed and starving for oxygen, the purest truth and the highest eloquence fall with little force.

We sleep in a small bed-room from which every breath of fresh air is excluded, because we believe night-air to be unhealthy,* and so we breathe its dozen hogsheads of air over and over again, and then wonder why we awaken in the morning so dull and unrefreshed! Return to our room after inhaling the fresh, morning air, and the fetid odor we meet on opening the door, is convincing proof how we have poisoned our lungs during the night.

Each room should be supplied with 2000 feet of fresh air per hour for every person it contains. Our ingenuity ought to find some way of doing this advantageously and pleasantly. A moiety of the care we devote to delicate articles of food, drink, and dress would abundantly meet this prime necessity of our bodies.

Open the windows a little at the top and the bottom. Put on plenty of clothing to keep warm by day and by night, and then let the inspiring oxygen come in as freely as God has given it. Pure air is the cheapest necessity and luxury of life. Let it not be the rarest!

School-room Ventilation.—Who, on going from the open air of a clear, bracing winter's day, into a crowded school-room, late in the session, has not noticed the disagreeable odor, and been for a moment nauseated, and half-stifled by the oppressive atmosphere! It is not strange. See how many causes here combine to pollute the air. If the room is heated by a stove, quantities of carbonic-oxide

---

* There is a singular prejudice against the night air. Yet, as Florence Nightingale aptly says, what other air can we breathe at night? We then have the choice between foul air within and pure air without. For, in large cities especially, the night air is far more wholesome than that of the day-time. To secure fresh air all night we must open the windows of our bed-room.

and carbonic-acid gases, as well as other products
of combustion, driven by downward drafts in the
flue, escape through seams and cracks and the
occasionally-opened door of the stove. In the case
of a furnace, the same effect is too often experienced and the odor of coal-gas is a common one,
especially when the fire is replenished. The insensible perspiration is more active in children than in
adults ; they, moreover, rush in with their clothing
saturated with the perspiration induced by their
sports ; so that, on the average, each pupil, during
school hours, loads the air with about half-a-pint of
aqueous vapor. The children come, oftentimes, from
homes that are close, ill-ventilated, and uncleanly ;
and frequently from sick-rooms, bringing in their
clothing the germs of disease. Some of the pupils
may even bear traces of illness, or have unsound
organs. and so their breath and exhalations be
poisonous.

In addition to all this, the air is filled with dust
brought in and kept astir by many busy feet ; by
ashes from the stove or furnace ; and especially by
chalk-dust. The modern method of teaching requires
a large amount of black-board work and the air of
the school-room is thus loaded with chalk-particles.
These collect in the nasal passages, and the upper
part of the larynx, and irritate the membrane, perhaps laying the foundation of catarrh.

The usual school-room atmosphere bears in the
pupils the natural fruit of frequent headaches, in-
attention, weariness, and stupor : but in the teacher
its frightful influence is most apparent. His labor
is severe, his worry of mind is constant, and, when

he finishes his day's work, he is generally too tired to take the required exercise. He consequently labors on with impaired health, or breaks down prematurely.

Instead of six hundred feet of space being allowed for each pupil, as perfect ventilation demands—the lowest estimate being 250 feet—often not over one hundred feet are afforded. Instead of 2000 cubic feet of fresh air for each pupil being supplied, and as much foul air removed every hour, as all physiologists assert is needed for perfect health, perhaps no means of ventilation at all are provided, and none is secured except what an occasionally-opened door, or the benevolent cracks and chinks in the building furnish the suffering lungs.*

How shall We Ventilate ?—The usual method of ventilation depends upon the fact that hot air is lighter than cold air, and so the cold air tends, by the force of gravity, to fall and compel the warm air to rise. Thus, if we open the door of a heated room, and hold a lighted candle first at the top, and then at the bottom, we can see, by the deflection of the flame, that there is a current of air setting outward at the top, and another setting inward at the bottom of the opening. A handkerchief held loosely, or the smoke of a smoldering match, in front of a fire-place will show a current of air passing up the

* Imagine fifty pupils put into a class-room thirty feet long, twenty-five feet wide, and ten feet high. This would generally be considered a very liberal provision. Such a room contains 7500 cubic feet of air. But it furnishes only 150 feet of space for each pupil. Allowing ten cubic feet of air per pupil each minute, in fifteen minutes after assembling, the entire atmosphere of the room is tainted, and unfit to be re-breathed. The demand of health is that at least 1500 cubic feet of pure air should be admitted into this room every minute, and as much be removed.

chimney ; this is caused by the difference of temperature between the air in the room and the outside
atmosphere. *Upon this difference of temperature, all
ordinary ventilation is based.** A proper treatment
of this subject and its practical applications, would
require a book by itself. There is room here for
only a few general statements and suggestions.

1. Two openings are always necessary to produce
a thorough change of air. (See *Chemistry*, p. 80.)
Put a lighted candle in a bottle. The flame will
soon be extinguished. The oxygen of the little air
in the bottle is burned out. and carbonic-acid has
taken its place. Now place over the mouth of the
bottle a lamp-chimney, and insert in the chimney a
strip of card-board, thus dividing the passage. On
relighting the candle, it will burn freely. The smoke
of a bit of smoldering paper will show that two
opposite currents of air are established, one setting
into the bottle, the other outward.

2. In the winter. when our school-rooms, churches,
public halls, etc., are heated artificially, ventilation
is comparatively easy if properly arranged.† The
required difference of temperature is kept up with
little difficulty. The fresh air admitted to the room
should then be heated either by a furnace, or by
passing over a stove, or through a coil of steam-

* Public buildings are sometimes ventilated by mechanical means, *i. e.*, immense
fans which are turned by machinery, and thus set the air in motion. Such methods
are, however, expensive, and rarely adopted, except where power is also used for
other purposes.

† For the escape of bad air, Dr. Bell suggests that an efficient foul-air shaft may
be fitted to the commonest of stoves by simply inclosing the stove-pipe in a jacket—
that is, in a pipe two or three inches greater in diameter. This should be braced
round the stove-pipe and left open at the end next the stove. At its entry into the
chimney, a perforated collar should separate it from the stove-pipe.

pipes. This cold air should always be taken directly from out-doors, and not from a cellar, or under a piazza where contamination is possible.

3. In order to remove the impure air, there should be ventilators provided at or near the floor, opening into air-shafts, or pipes leading upward through the roof, with proper orifices at the top. These ventilating-pipes should be heated artificially so as to produce a draft. They may form one of the flues of a chimney in which there is a constant fire ; or be carried upward in a large flue through the center of which runs the smoke-pipe of the furnace or stove ; * or the ventilating-pipe be itself conveyed through the center of the larger chimney-flue. If the register for hot air be on the floor at one side of the room, two or more ventilators may be placed near the floor on the opposite side. The warm air will thus make the complete circuit of the room, and thoroughly warm it before passing out.

If the ventilating-shaft be not heated artificially, the ventilator must be placed at the top of the room in order that the hot air may escape through it, thus . producing an upward draft. But the objection to this method is that it allows the warmer air to escape, while economy requires that the cooler air at the bottom of the room should be removed and

* This plan has been adopted in the new school-buildings of Elmira, N. Y. The older buildings were provided with ventilating-pipes, not heated artificially, and hence of no service. These pipes are rendered effective, however, by conducting them into a small room in the garret, heated by a coal-stove. From this room, a large exit-pipe leads to the roof, where it terminates in an Emerson's ventilator. So strong a draft is thus established that throughout the building air is taken from the floors, and consequently the cooler portion of the rooms, at a velocity of three to five feet per second, or 180 to 300 cubic feet per minute for each square foot of flue-opening. In perpendicular flues, heated throughout with a smoke-flue from the furnace, ten feet per second is attained.

the warm air be made to descend, thus securing uniformity of temperature.

4. In the summer, ventilation may be commonly provided for by opening windows *at the top and the bottom*, on the sheltered side of the building, so as to avoid drafts of air injurious to the occupants. On a dull, still, hot day, when there is little difference of temperature between the inner and the outer air, ventilation can be secured only by having a fire provided in the ventilating-shaft; this, by exhausting the air from the room, will cause a fresh current to pour in through the open windows. At recess, all the children should, if the weather permit, be sent out-doors, to allow their clothing to be exposed to the purifying influence of the open air, and the windows to be thrown wide open, to ventilate the room thoroughly. In bad weather, rapid marching or calisthenic exercises will furnish exercise, and also permit the airing of the room.

5. The school and the church are the centers for spreading contagious diseases. The former is especially dangerous, and therefore great pains should be taken to exclude pupils attacked by or recovering from diphtheria, scarlet-fever, whooping-cough, etc., and even those who live in houses where such sickness exists.

6. In our houses,* open fire-places are efficient

* The air of our homes is often contaminated by decaying vegetables and other filth in the cellar; by bad air drawn up into the cellar from the soil, by the powerful draughts that our fires create; by defective gas and waste-pipes that let the foul air from cesspool or sewer spread through the house; and by piles of refuse, or puddles of slops emptied at the back-door. While the water in our wells, or in streams that supply our towns and cities, receives too often the drainage from out-houses and barn-yards, and so introduces into our systems, in the liquid, and thus easily-assimilated form, the most dangerous poisons. The question of sanitary precautions is therefore one that presses upon every thoughtful mind, and demands constant attention.

ventilators, and they should never be closed for any cause. Fresh air admitted by a hot-air register and impure air passed out by a chimney, form a simple and thorough system. Our sleeping-apartments demand especial care. As soon as the occupants leave the room, the bed-clothes should be removed, and laid on the backs of chairs to air ; the bed be shaken up ; and the windows thrown wide. In the summer, the windows may be closed before the sun is high ; the house is then left filled with the cool morning air. In damp and cold weather, a fire should be lighted in sleeping-apartments, particularly if used by children or delicate persons, to dry the bed-clothing, and also to prevent a chill on the part of the occupants. It is not necessary to go shivering to bed in order to harden one's constitution.

Wonders of Respiration.—The perfection of the organs of respiration challenges our admiration. So delicate are they that the least pressure would cause exquisite pain, yet tons of air surge to and fro through their intricate passages, and bathe their innermost cells. We yearly perform at least 7,000,000 acts of breathing, inhaling 100,000 cubic feet of air, and purifying over 3,500 tons of blood. This gigantic process goes on constantly, never wearies or worries us, and we wonder at it only when science reveals to us its magnitude. In addition, by a wise economy, the process of respiration is made to subserve a second use no less important, and the air we exhale, passing through the organs of voice, is transformed into prayers of faith, songs of hope, and words of social cheer.

Diseases, etc.—1. CONSTRICTION OF THE LUNGS is
produced by tight clothing. The ribs are thus forced

Fig. 33.

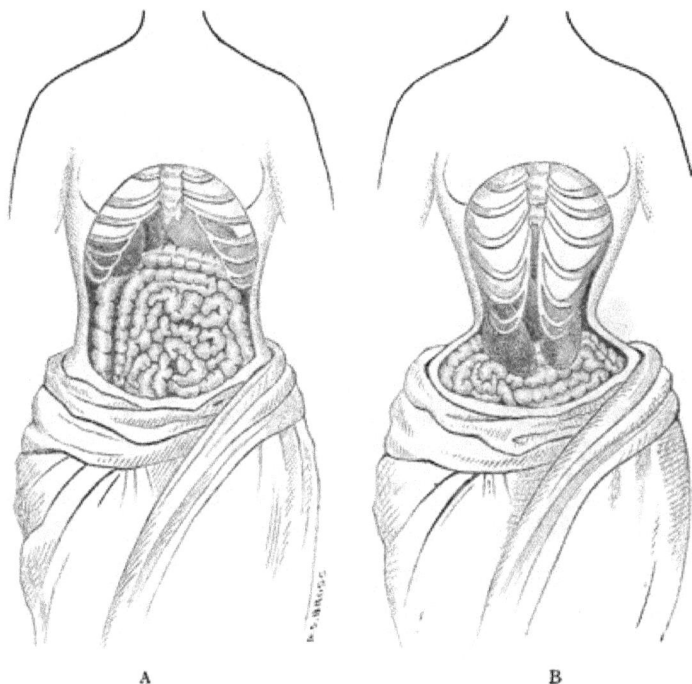

A   B

A, *the natural position of the internal organs.* B. *when deformed by tight lacing.*
MARSHALL *says that the liver and the stomach have, in this way, been forced downward almost as low as the pelvis.*

inward, the size of the chest is diminished, and the
amount of inhaled air decreased. Stiff clothing, and
especially a garment that will not admit of a full
breath without inconvenience, will prevent that free
movement of the ribs so essential to health. Any
infraction of the laws of respiration, even though it
be fashionable, will result in diminished vitality and

vigor, and will be fearfully punished by sickness and weakness through the whole life.

2. BRONCHITIS is an inflammation (see Inflammation) of the mucous membrane of the bronchial tubes. It is accompanied by an increased secretion of mucus, and consequent coughing.

3. PLEURISY is an inflammation of the pleura. It is sometimes caused by an injury to the ribs, and results in a secretion of water within the membrane.

4. PNEUMONIA (*pneuma*, breath) is an inflammation of the lungs, affecting chiefly the air-cells.

5. CONSUMPTION is a disease which destroys the substance of the lungs. Like other lung difficulties, it is caused largely by a want of pure air, a liberal supply of which is the best treatment that can be prescribed for it.*

6. ASPHYXIA (as-fix'-i-a). When a person is drowned, strangled, or choked in any way, what is called asphyxia occurs. The face turns black ; the veins become turgid ; insensibility and often convulsions ensue. If relief is not secured within a few minutes, death will be inevitable.† (See Appendix.)

7. DIPHTHERIA (*diphthera*, a membrane) is a kind of sore-throat, in which matter exudes from the mucous membrane. This stiffens into a peculiar white substance, patches of which may be seen in

---

* " If I were seriously ill of consumption, I would live out-doors day and night, except in rainy weather or mid-winter ; then I would sleep in an unplastered log-house. Physic has no nutriment, gaspings for air cannot cure you, monkey capers in a gymnasium cannot cure you, and stimulants cannot cure you. What consumptives want is pure air, not physic—pure air, not medicated air—plenty of meat and plenty of bread."—*Dr. Marshall Hall.*

† The lack of oxygen, and the presence of carbonic-acid gas, are the combined causes. Oxygen starvation and carbonic-acid poisoning, each fatal in itself, work together to destroy life.

the back part of the mouth. Fever and debility accompany this disease, which is so sudden and insidious in its advances as to be exceedingly dreaded.

8. CROUP, which often attacks young children, is an inflammation of the mucous membrane of the larynx and trachea. It is commonly preceded by a cold. The child sneezes, coughs, and is hoarse, but the attack frequently comes on suddenly, and usually in the night. It is accompanied by a peculiar "brassy," ringing cough, which, once heard, can never be mistaken. It may prove fatal within a few hours. (See Appendix.)

9. STAMMERING depends, not on defects of the muscles, but on a want of due control of the mind. When a stammerer is not too conscious of his lack, and tries to form his words slowly, he speaks plainly, and may sing well, for then his words must come in time. Many persons who stutter in common conversation can talk with much fluency when making a speech. The stammerer should find out his peculiar difficulty, and overcome it by exercise, and especially by speaking only after a full inspiration.

PRACTICAL QUESTIONS

1. What is the philosophy of "the change of voice" in a boy?

2. Why can we see our breath on a frosty morning?

3. When a law of health and a law of fashion conflict, which should we obey?

4. If we use a "bunk" bed, should we pack away the clothes when we first rise in the morning?

5. Why should a clothes-press be well ventilated?

6. Should the weight of our clothing hang from the waist, or the shoulder?

7. Describe the effects of living in an overheated room?

8. What habits impair the power of the lungs?

9. For full, easy breathing in singing, should we use the diaphragm and lower ribs, or the upper ribs alone?

10. Why is it better to breathe through the nose than the mouth?

11. Why should not a speaker talk while returning home on a cold night after a lecture?

12. What part of the body needs the loosest clothing?

13. What part needs the warmest?

14. Why is a "spare bed" generally unhealthful?

15 Is there any good in sighing?

16. Should a hat be thoroughly ventilated? How?

17. Why do the lungs of people who live in cities become of a gray color?

18. How would you convince a person that a bed-room should be aired?

19. What persons are most liable to scrofula, consumption, etc.?

20. If a person is plunged under water will it enter his lungs?

21. Are bed-curtains healthful?

22. Why do some people take "short breaths" after a meal?

23. What is the special value of public parks?

24. Can a person become used to bad air, so that it will not injure him?

25. Why do we gape when we are sleepy?

26. Is a fashionable waist a model of art in sculpture or painting?

27. Should a fire-place be closed?*

28. Why does embarrassment or fright cause a stammerer to stutter still more painfully?

---

* "Thousand of lives would be saved if all fire-places were kept open. If you are so fortunate as to have a fire-place in your room, paint it when not in use, put a bouquet of fresh flowers in it every morning, if you please, or do anything to make it attractive, but never *close it;* better use the fire-boards for kindling-wood. It would be scarcely less absurd to take a piece of elegantly-tinted court-plaster and stop up the nose, trusting to the accidental opening and shutting of the mouth for fresh air, because you thought it spoiled the looks of your face to have two such great, ugly holes in it, than to stop your fire-place with elegantly-tinted paper, or a Japanese fan, because it looks better."—*Leeds.*

29. In the organs of voice, what parts have somewhat the same effect as the case of a violin and the sounding-board of a piano?

30. Why should we be careful not to "take the breath of a sick person"?

31. What special care should be taken with regard to keeping a cellar clean?

32. How is the air strained as it passes into the lungs?

33. Can one really "draw the air into his lungs?"

34. How often do we breathe?

35. Describe some approved method of ventilation.

36. What is at once the floor of the chest and the roof of the abdomen?

37. What would you do in a case of apparent death by drowning, or by coal-gas? (See Appendix.)

38. What would you do in a case of croup, while the doctor was coming? (See Appendix.)

39. How would you treat a severe burn? (See Appendix.)

40. Describe the various ways in which the water in a well is liable to become unwholesome.

V.

CIRCULATION.

" No rest this throbbing slave may ask,
For ever quivering o'er his task,
While far and wide a crimson jet
Leaps forth to fill the woven net,
Which in unnumber'd crossing tides
The flood of burning life divides,
Then, kindling each decaying part,
Creeps back to find the throbbing heart."

HOLMES.

BLACKBOARD ANALYSIS.

THE CIRCULATION.

1. THE BLOOD
   1. Its Composition.
   2. Its Uses.
   3. Transfusion.
   4. Coagulation.

2. ORGANS OF THE CIRCULATION.
   1. The Heart.
      1. Description.
      2. Movements.
      3. Auricles and Ventricles.
      4. The Valves.
         a. Need of.
         b. Tricuspid and Bi-cuspid.
         c. The Strengthening of the Valves.
         d. Semi-lunar Valves.
   2. The Arteries.
      1. Description.
      2. The Arterial System.
      3. The Pulse.
   3. The Veins.
      1. General Description.
      2. Valves.
   4. The Capillaries.
      1. Description.
      2. Use.
      3. Under the Microscope.

3. THE CIRCULATION
   1. The Lesser.
   2. The Greater.
   3. The Velocity of the Blood.

4. THE HEAT OF THE BODY.
   1. Distribution.
   2. Regulation.

5. LIFE BY DEATH.

6. CHANGE OF OUR BODIES.

7. THE THREE VITAL ORGANS.

8. WONDERS OF THE HEART.

9. THE LYMPHATIC CIRCULATION.
   1. Description.
   2. The Glands.
   3. The Lymph.
   4. The Office of the Lymphatics.

10. DISEASES
    1. Congestion.
    2. Inflammation.
    3. Bleeding.
    4. Scrofula.
    5. A Cold.
    6. Catarrh.

11. ALCOHOLIC DRINKS AND NARCOTICS.
    1. Effect of Alcohol upon the Circulation.
    2. "   "   "   "   "   Heart.
    3. "   "   "   "   "   Membrane.
    4. "   "   "   "   "   Blood.
    5. "   "   "   "   "   Lungs.

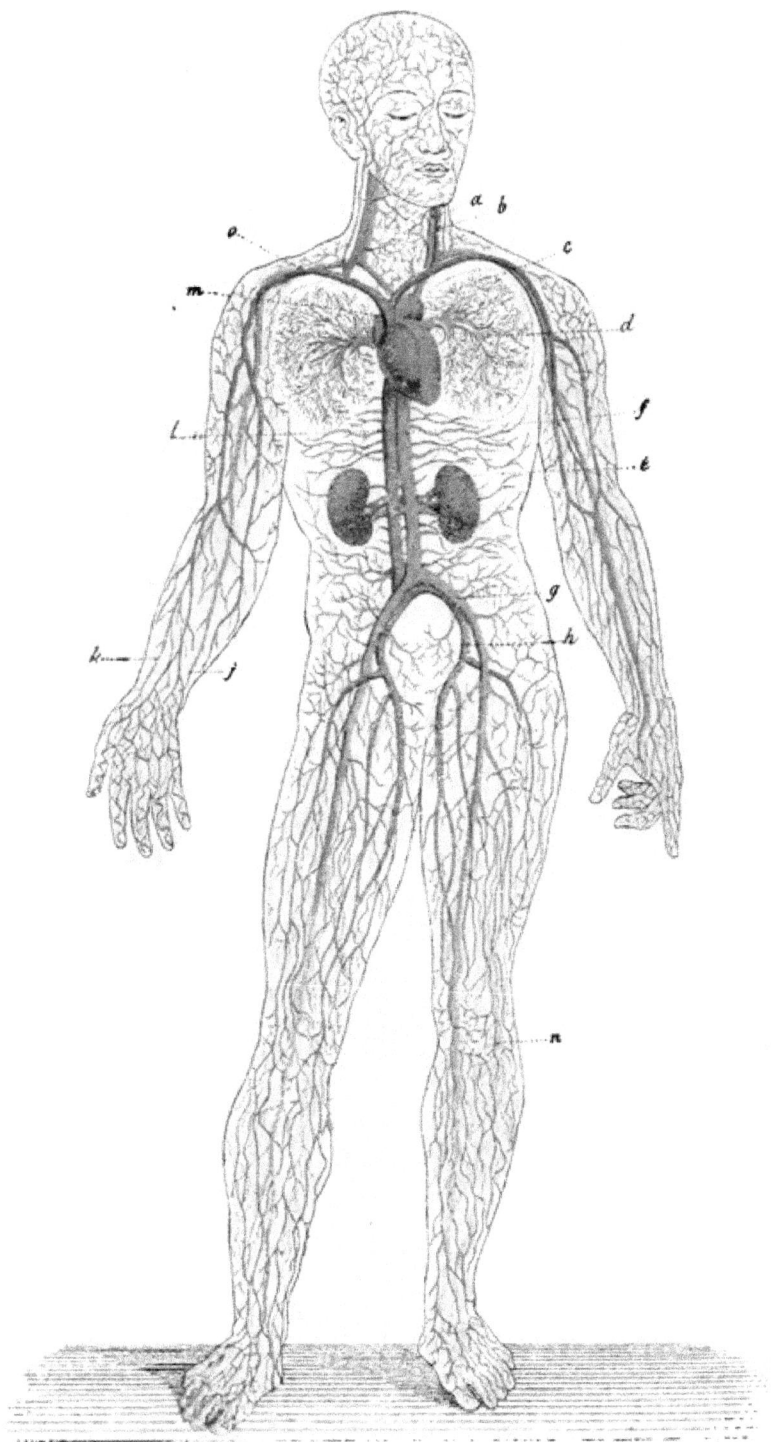

a   b
o
c
m
d
f
t
l
g
h
k
j
n

THE CIRCULATION.

T HE **Organs of the Circulation** are the *heart*, the *arteries*, the *veins*, and the *capillaries*.

The Blood is the liquid by means of which the circulation is effected. It permeates every part of

Fig. 35.

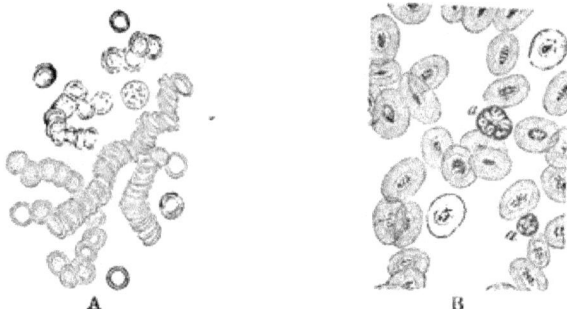

A B

A, *corpuscles of human blood, highly magnified* ; B, *corpuscles in the blood of an animal (a non-mammal).*

the body, except the cuticle, nails, hair, etc. The average quantity in each person is about eighteen pounds.* It is composed of a thin, colorless liquid, the *plasma*, filled with red disks or cells,† so small

---

* It is difficult to estimate the exact amount, and therefore authorities disagree. Foster places it at about one-thirteenth of the body-weight.

† "There is also one white globular cell to every three or four hundred red ones. The blood is no more red than the water of a stream would be if you were to fill it with little red fishes. Suppose the fishes to be very, very small—as small as a grain

that about 3,500 placed side by side would measure only an inch, and it would take 16,000 laid flatwise upon one another to make a column of that height. Under the microscope, they are found to be rounded at the edge and concave on both sides.* They have a tendency to collect in piles like rolls of coin. The size and shape vary in the blood of different animals.† Disks are continually forming in the blood, and as constantly dying—20,000,000 at every breath (*Draper*).

The plasma also contains fibrin,‡ albumen—which

of sand—and closely crowded together through the whole depth of the stream ; the water would look quite red. would it not ? And this is the way in which blood looks red—only observe one thing ; a grain of sand is a mountain in comparison with the little red fishes in the blood.   If I were to tell you they measured about $\frac{1}{3000}$ of an inch in diameter, you would not be much wiser ; so I prefer saying (by way of giving you a more perfect idea of their minuteness) that there would be about a million in such a drop of blood as would hang on the point of a needle. I say so on the authority of a scientific microscopist—M. Bouillet.   Not that he has ever counted them, as you may suppose, any more than I have done ; but this is as near an approach as can be made by calculation to the size of $\frac{1}{3000}$ part of an inch in diameter."
—*Jean Mace.*

* By pricking the end of the finger with a needle, we can obtain a drop for examination.   Place it on the slide, cover with a glass. and put it at once under the microscope.   The red disks will be seen to group themselves in rows, while the white disks will seem to draw apart, and to be constantly changing their form.   After a gradual evaporation, the crystals (Fig. 36) may be seen.   In animals, they have various, though distinctive forms.

† Authorities differ greatly in their estimate of the size of the disks (corpuscles) in human blood.   The fact is that the size varies in different persons. probably also in the same individual.   Many of the best microscopists therefore hesitate to state whether a particular specimen of blood belonged to a human being or to an animal.   Others claim that they can distinguish with accuracy.   Evidently, the question is one of great uncertainty.   The following statement of the size of the cells in different animals is taken from Gulliver's tables : Cat. $\frac{1}{3157}$ of an inch in diameter ; whale, $\frac{1}{3100}$; mouse, $\frac{1}{3217}$; hog, $\frac{1}{4230}$; camel, $\frac{1}{5305}$; sheep, $\frac{1}{5300}$; horse, $\frac{1}{4600}$; Virginia deer, $\frac{1}{5412}$: dog-faced baboon, $\frac{1}{3187}$; brown baboon, $\frac{1}{3463}$; red monkey, $\frac{1}{3305}$; black monkey, $\frac{1}{3578}$.

‡ It is usual to say that fibrin is contained in the blood.   It probably does not exist as such. but there are present in the blood certain substances known as *para-globulin* and *fibrinogin*, which, by the action of the third substance, *fibria-ferment*, under certain circumstances. form(fibrin and so cause coagulation.   The exact nature of the process by which fibrin is produced is not understood.   See *Foster's Text Book of Physiology*, p. 22.

Fig. 36.

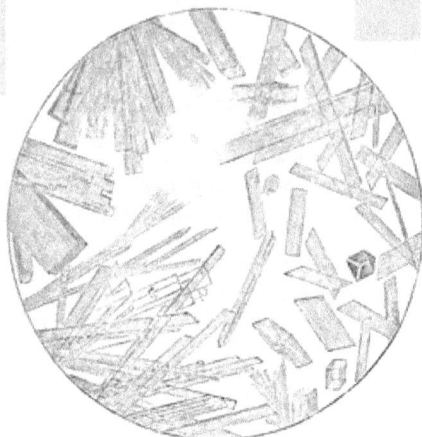

Blood Crystals.

is found nearly pure in the white of an egg — and also various mineral substances, as iron,* lime, magnesia, phosphorus, potash, etc.

Uses of the Blood.—The blood has been called "liquid flesh;" but it is more than that, since it contains the materials for making every organ. The plasma is rich in mineral matter for the bones, and in albumen for the muscles. The red disks are the air-cells of the blood. They contain the oxygen so essential to every operation of life. Wherever there is work to be done or repairs to be made, there the oxygen is needed. It stimulates to action, and tears down all that is worn out. In this process, it combines with and actually burns out parts of the muscles and other tissues, as wood is burned in the stove.† The blood, now foul with the burned matter.

---

* Enough iron has been found in the ashes of a burned body to form a mourning ring.

† For the sake of simplicity, perhaps to conceal our own ignorance, we call this process "burning." The simile of a fire is good so far as it goes. But as to the real

the refuse of this fire, is caught up by the circulation, and whirled back to the lungs, where it is purified, and again sent bounding on its way.

There are, then, two different kinds of the blood in the body : the red or arterial, and the dark or venous.

Transfusion.—As the blood is really the "vital fluid," efforts have been made to restore the feeble by infusing healthy blood into their veins. If blood be drawn from an animal until it is seemingly dead, and then that from another animal be injected into its veins, its vitality will be restored.* This practice became quite common in the seventeenth century. The operation was even tried on human beings, and

nature of the change which the physiologist briefly terms "oxidation," we know nothing. This much only can be asserted positively. A stream of oxygen is carried by the blood to the muscles (in fact to every tissue in the body), while, from the muscles the blood carries away a stream of carbonic-acid and water. But what takes place in the muscles, when and what chemical change occurs, no one can tell. We see the first and the last stage. We know that contraction of the muscles somehow comes about, oxygen disappears, carbonic-acid appears, energy is released, and force is exhibited as motion, heat, and electricity. But the intermediate step is hidden.

There are certain theories, however, advanced that are worth considering. Some physiologists hold that the muscle has the power of taking up the oxygen from the haemoglobin (a body that comprises ninety per cent. of the red corpuscles when dried, and is the oxygen-carrier of the blood), and fixing it, as well as the raw material (food) furnished by the blood, thus forming a true contractile substance. The breaking-down or decomposition of this contractile substance in the muscle, sets free its potential energy. The process is gentle so long as the muscle is at rest, but becomes excessive and violent when contraction occurs. (See Foster's Physiology, p. 118.) It is also believed by some that the chemical change in the muscle partakes of a fermentive character; that, under the influence of the proper ferments, the substances break up into other and simpler products, thus setting free heat and force ; and that this chemical change is followed by a secondary oxidation by the oxygen in the arterial blood, thereby forming carbonic-acid and water, as in all putrefactive processes. But these and other views are not as yet fully understood ; while they utterly fail to tell us how a collection of simple cells, filled merely with a semi-fluid mass of matter, can contract and set free muscular power. The commonness of this act hides from us its wonderful nature. But here, hidden in the cell—Nature's tiny laboratory—lies the mystery of life. Before its closed door we ponder in vain, confessing the unskillfulness of our labor, and fearing all the while lest the *Secret of the Cell* will always elude our search.

* Brown-Séquard tells of a curious instance in which the blood of a living dog was transferred into one just dead. The animal rose on its feet and wagged its tail, but died a second time twelve and one-half hours afterward.

the most extravagant hopes were entertained. A maniac was restored to reason by the blood of a calf. But many fatal accidents occurring, it was forbidden by law, and soon fell into disuse. It has, however, been successfully practiced in several cases within the last few years, and is a method still in repute for saving life.

Coagulation.—When blood is exposed to the air, it coagulates. This is caused by the solidifying of the fibrin, which, entangling the disks, forms the "clot." The remaining clear, yellow liquid is the *serum.* The value of this peculiar property of the blood can hardly be over-estimated. The coagulation soon checks all ordinary cases of bleeding.* When a wound is made, and bleeding commences, the fibrin forms a temporary plug, as it were, which is absorbed when the healing process is finished. Thus we see how a Divine foresight has provided not only for the ordinary wants of the body, but also for the accidents to which it is liable.†

The Heart is the engine which propels the blood. It is a hollow, pear-shaped muscle, about the size of the fist. It hangs, point downward, just to the left of the center of the chest. (See Fig. 31.) It is enclosed in a loose sac of serous membrane,‡ called the

---

* In the case of the lower animals, which have no means of stopping hemorrhages as we have, the coagulation is far more rapid.

† The fibrin is not an essential ingredient of the blood. All the functions of life are regularly performed in people whose blood lacks fibrin; and, in cases of transfusion, where blood deprived of its fibrin was used, the vivifying influence seemed to be the same. Its office, therefore, must mainly be to stanch any hemorrhage which may occur.—*Flint.*

‡ The mucous membrane lines the open cavities of the body; the serous, the closed. The pericardium is a sac composed of two layers—a fibrous membrane on the outside, and a serous one on the inside. The latter covers the external surface

pericardium (*peri*, about; and *kardia*, the heart),
This secretes a lubricating fluid, and is smooth as
satin.

The Movements of the Heart consist of an alter-

Fig. 37.

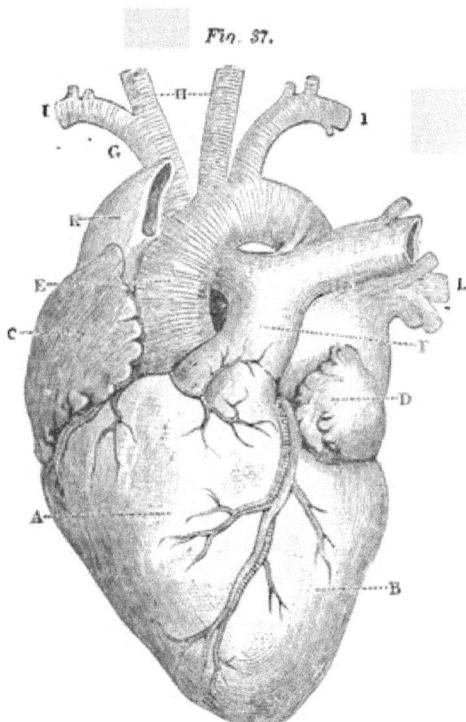

The Heart. A, *the right ventricle;* B, *the left ventricle;* C, *the right auricle*
D, *the left auricle.*

nate contraction and expansion. The former is
called the *sys'to-le,* and the latter the *di-as'to-le.*

of the heart, and is reflected back upon itself in order to form, like all the membranes
of this nature, a sac without an opening. The heart is thus covered by the pericar-
dial sac, but not contained inside its cavity. A correct idea may be formed of the
disposition of the pericardium around the heart by recalling a very common and very
convenient, though now discarded head-dress, the cotton night-cap. The pericardium
encloses the heart exactly as this cap covered our forefathers' heads.— *Wonders of the
Human Body.*

During the diastole, the blood flows into the heart, to be expelled by the systole. The alternation of these movements constitutes the beating of the heart which we hear so distinctly between the fifth and sixth ribs.*

Fig. 38

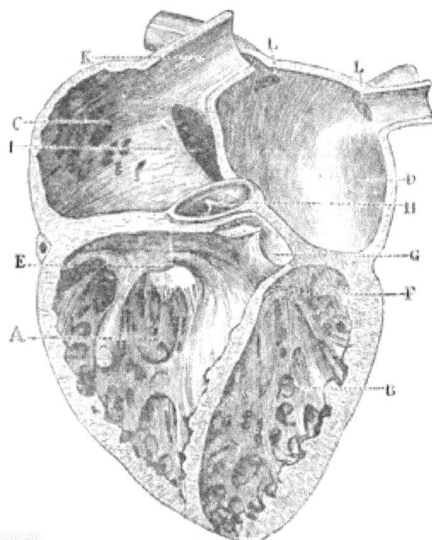

Chambers of the Heart. A, *right ventricle;* B, *left ventricle;* C, *right auricle,* D. *left auricle:* E, *tricuspid valve;* F, *bicuspid valve;* G, *semi-lunar valves;* H, *valve of the aorta;* I, *inferior vena cava;* K, *superior vena cava;* L, L, *pulmonary veins.*

The Auricles and Ventricles.—The heart is divided into four chambers. In an adult, each holds about a wine-glassful. The upper ones, from appendages on the outside resembling the ears of a dog, are

---

* "Two sounds are heard if we put our ear over the heart,—the first and longer as the blood is leaving the organ, the second as it falls into the pockets of the two arteries, and the valves then striking together cause it. The first sound is mainly the noise made by the muscular tissue. During the first, the two ventricles contract; during the second, the two auricles do so. The hand may feel the heart striking the ribs as it contracts,—a feeling called the impulse, or, if quicker and stronger than usual, palpitation. This is not always a sign of disease, but in hypochondriacs is often an effect of the mind on the nerves of the heart."—*Mapother.*

called *auricles* (*aures*, ears) ; the lower ones are termed *ventricles*. The auricle and ventricle on each side communicate with each other, but the right and left halves of the heart are entirely distinct, and perform different offices. The left side propels the red blood : and the right, the dark.

The auricles are merely reservoirs to receive the blood, (the left auricle, as it filters in bright and pure from the lungs ; the right, as it returns dark and foul from the tour of the body), and to furnish it to the ventricles as they need. Their work being so light, their walls are comparatively thin and weak. On the other hand. the ventricles force the blood, (the left, to all parts of the body ; the right, to the lungs), and are, therefore, made very strong. As the left ventricle drives the blood so much further than the right, it is correspondingly thicker and stronger.

Need of Valves in the Heart.—As the auricles do not need to contract with much force simply to empty their contents into the ventricles below them, there is no demand for any special contrivance to prevent the blood from setting back the wrong way. Indeed. it would naturally run down into the ventricle, which is at that moment open to receive it. But. when the strong ventricles contract, especially the left one which must drive the blood to the extremities, some arrangement is necessary to prevent its escaping into the auricle again. Besides, when they expand. the "suction power" would tend to draw back again from the arteries all the blood just forced out. This difficulty is obviated by

Fig. 39.

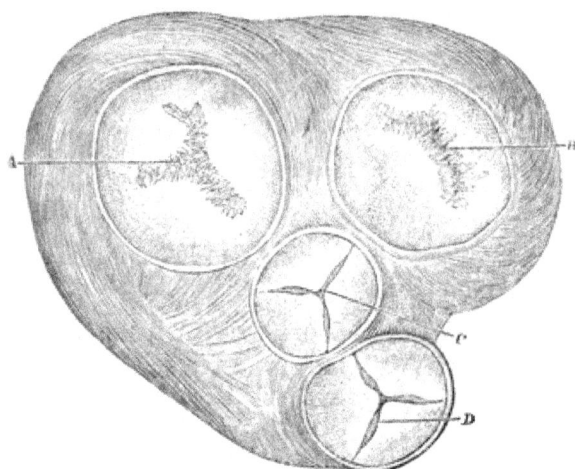

Diagram showing the peculiar Fibrous Structure of the Heart and the Shape of the Valves. A, tricuspid valve ; B, bicuspid valve ; C, semi-lunar valves of the aorta ; D, semi-lunar valves of the pulmonary artery.

means of little doors, or valves, which will not let it go the wrong way.[*]

The Tricuspid and Bicuspid Valves. — At the opening into the right ventricle, is a valve consisting of three folds or flaps of membrane, whence it is called the *tri-cuspid* valve (*tri*, three ; and *cuspides*, points), and in the left ventricle, one containing two flaps, and named the *bi-cuspid* valve. These hang so loosely as to oppose no resistance to the passage of the blood into the ventricles ; but, if any attempts to go the other way, it gets between the flaps and the walls of the heart, and, driving them outward. closes the orifice.

[*] The heart of an ox or a sheep may be used to show the chambers and valves. The aorta should be cut as far as possible from the heart, and then by pumping in water the perfection of these valves will be finely exhibited. Cutting the heart across near the middle will show the greater thickness of the left ventricle.

These Flaps are Strengthened like sails by slender cords, which prevent their being pressed back through the opening. If the cords were attached directly to the walls of the heart, they would be loosened in the systole, and so become useless when most needed. They are, therefore, fastened to little muscular pillars projecting from the sides of the ventricle; when that contracts the pillars contract also, and thus the cords are held tight.

The Semi-lunar Valves.—In the passages outward from the ventricles, are valves, called from their peculiar half-moon shape *semi-lunar* valves (*semi*, half; *Luna*, Moon). Each consists of three little pocket-shaped folds of membrane, with their openings in the direction which the blood is to take. When it sets back, they fill, and, swelling out, close the passage (Fig. 40).

The Arteries* are the tube-like canals which convey the blood *from* the heart. They carry the red blood (see note, p. 118). They are composed of an elastic tissue, which yields at every throb of the heart, and then slowly contracting again, keeps up the motion of the blood until the next systole. The elasticity of the arteries acts like the air-chamber of a fire-engine, which converts the intermittent jerks of the brakes or pump into the steady stream of the hose-nozzle.

The arteries sometimes communicate by means of branches or by meshes of loops, so that if the blood be blocked in one, it can pass round through another,

* *Aer*, air: and *terco*, I contain—so named because after death they contain air only, and hence the ancients supposed them to be air-tubes leading through the body.

and so get by the obstacle.* When an artery penetrates a muscle, it is often protected by a sheath or by fibrous rings, which prevent its being pulled out of place or compressed by the play of the muscles.

The arteries are generally located as far as possible beneath the surface, out of harm's way, and hence are found closely hugging the bones or creeping through safe passages provided for them. They are generally nearly straight, and take the shortest routes to the parts which they are to supply with blood.

The Arterial System starts from the left ventricle by a single trunk—the *aorta*—which, after giving off branches to the head, sweeps back of the chest with a bold curve—the *arch of the aorta* (c, Fig. 34) —and thence runs downward (f), dividing and subdividing, like a tree, into numberless branches, which, at last, penetrate every nook and corner of the body.

The Pulse.—At the wrist (k, radial artery) and on the temple (temporal artery) we can feel the expansion of the artery by each little wave of blood set in motion by the contraction of the heart. In health, there are about seventy-two† pulsations per minute. They increase with excitement or inflammation,

---

* This occurs especially about the joints, where it serves to maintain the circulation during the bending of a limb, or when the main artery is obstructed by disease or injury, or has been tied by the surgeon. In the last case, the small adjacent arteries gradually enlarge, and form what is called a collateral circulation.

† This number varies much with age, sex, and individuals. Napoleon's pulse is said to have been only 40, while it is not infrequent to find a healthy pulse at 100 or over. Shame makes the heart send more blood to the blushing cheek, and fear almost stops it. The will cannot check the heart. There is said, however, to have been a notable exception to this in the case of one Col. Townsend, of Dublin, who, after having succeeded several times in stopping the pulsation, at last lost his life in the act.

weaken with loss of vigor, and are modified by nearly every disease. The physician, therefore, finds the pulse a good index of the state of the system and the character of the disorder.

The Veins are the tube-like canals which convey the blood *to* the heart.* They carry the dark or venous blood (note, p. 118). As they do not receive the direct impulse of the heart, their walls are made much thinner and less elastic than those of the arteries. At first small, they increase in size and diminish in number as they gradually pour into one another, like tiny rills collecting to form two rivers, the vena cava ascending and the vena cava descending (*l, m,* Fig. 34), which empty into the right auricle.

Some of the veins creep along under the skin, where they can be seen, as in the back of the hand; while others accompany the arteries, some of which have two or more of these companions.

Valves similar in construction to those already described (the semi-lunar valves of the heart, page 112) are placed at convenient intervals, in order to guide the blood in its course, and prevent its setting backward.† We can easily

---

* There is one exception to the general course of the veins. The *portal* vein carries the blood from the digestive organs to the liver, where it is acted upon, thence poured into the ascending vena cava, and goes back to the heart.

† "Too much standing, or tight elastics, often swell and spoil the valves of the veins in the leg: they then become *varicose,* or permanently enlarged, and, if they burst, the bleeding may be profuse and even dangerous. Raising the leg and pressing the finger on the bleeding spot will stay it. Walking does not encourage this disease, for the muscles force on the venous blood. Clerks who are subject to varicose veins should have seats behind the counters where they may rest when not actually employed. A deep breath helps the flow in the veins, and a wound may suck in air with fatal effect. A maimed horse is most mercifully killed by blowing a bubble of air into the veins of his neck. As the pressure deep in the sea would burst valves, there are none in the whale, and hence a small wound by the harpoon causes him to bleed to death."—*Mapother.*

examine the working of these valves. On baring the arm, blue veins may be seen running along the arm toward the hand. Their diameter is tolerably even, and they gradually decrease in size. If now the finger be pressed on the upper part of one of these veins, and then passed downward so as to

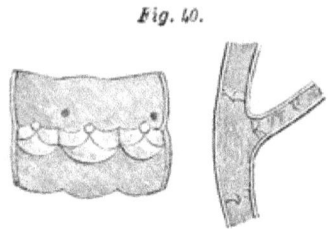

Fig. 40.

Valves of the Veins.

drive its blood backward, swellings like little knots will make their appearance. Each of these marks the location of a valve, which is closed by the blood we push before our finger. Remove the pressure, and the valve will swing open, the blood set forward, and the vein collapse to its former size.

The Capillaries (*capillus*, a hair) form a fine network of tubes, connecting the ends of the arteries with the veins. They blend, however, with the extremities of these two systems, so that it is not easy to tell just where an artery ends and a vein begins. So closely are they placed, that we cannot prick the flesh with a needle without injuring, perhaps, hundreds of them. The air-cells of the blood deposit there their oxygen, and receive carbonic acid, while in the delicate capillaries of the lungs* they give up their load of carbonic acid in exchange for oxygen.

If, by means of a microscope, we examine the transparent web of a frog's foot, we can trace the

* The capillary tubes are there so fine that the disks of the blood have to go one by one, and are sadly squeezed at that. However, their elasticity enables them to resume their old shape as soon as they have escaped from this labyrinth.

Fig. 41.

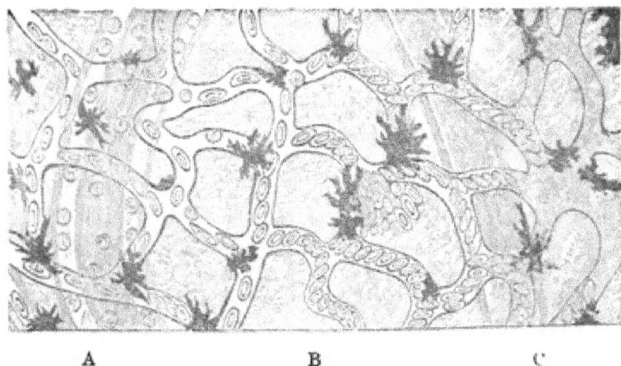

A   B   C

Circulation of the Blood in the Web of a Frog's Foot, highly magnified. A. an artery; B. capillaries crowded with disks, owing to a rupture just above, where the disks are jammed into an adjacent mesh; C, a deeper vein; the black spots are pigment cells.

route of the blood.* It is an experiment of wonderful interest. The crimson stream, propelled by the heart, rushes through the arteries, until it reaches the intricate meshes of the capillaries. Here it breaks into a thousand tiny rills. We can see the disks winding in single file through the devious passages, darting hither and thither, now pausing, swaying to and fro with an uncertain motion, and anon dashing ahead. until. at last. gathered in the veins, the blood sets steadily back on its return to the heart.

The Circulation † consists of two parts—the *lesser*, and the *greater*.

* With small splints and twine, a frog's foot can be easily stretched and tied so that the transparent web can be placed on the table of the microscope.

† The circulation of the blood was discovered by Harvey in 1619. For several years, he did not dare to publish his belief. When it became known, he was bitterly persecuted, and his practice as a physician greatly decreased in consequence. He lived, however, to see his theory universally adopted. and his name honored. Harvey is said to have declared that no man over forty years of age accepted his views.

1. THE LESSER CIRCULATION. — The dark blood from the veins collects in the right auricle, and, going through the tricuspid valve, empties into the

Fig. 42.

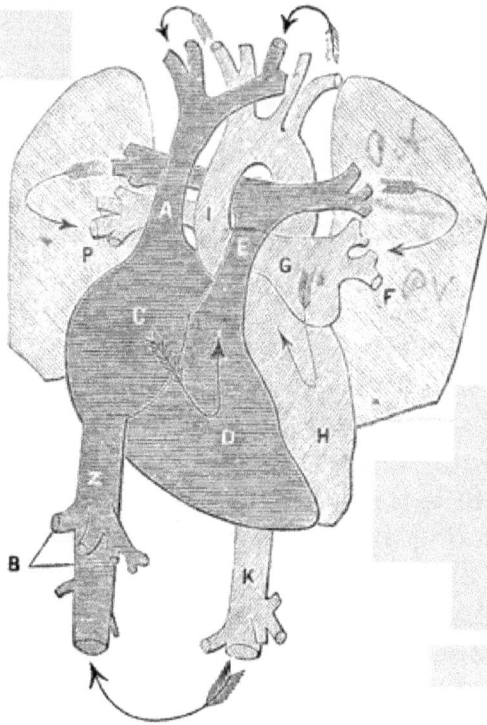

Diagram illustrating the Circulation of the blood.—MARSHALL. A, *vena cava descending (superior)*; Z. *vena cava ascending (inferior)*; C, *right auricle*; D, *right ventricle*; E, *pulmonary artery*; F P, *lungs and pulmonary veins*; G, *left auricle*; H, *left ventricle*; I, K, *aorta.*

right ventricle. Thence it is driven past the semi-lunar valves, through the pulmonary artery, to the lungs. After circulating through the fine capillaries of the air-cells, it is returned, bright and red, through the four pulmonary veins,* to the left auricle.

2. THE GREATER CIRCULATION.—From the left auricle, the blood is forced past the bicuspid valve to the left ventricle ; thence it is driven through the semi-lunar valves into the great aorta, the main trunk of the arterial system. Passing through the arteries, capillaries, and veins, it returns through the venæ cavæ, ascending and descending, gathers again in the right auricle, and so completes the "grand round" of the body. Both of these circulations are going on constantly, as the two auricles contract, and the two ventricles expand simultaneously, and *vice versa*.

The Velocity of the Blood varies so much in different parts of the body, and is influenced by so many circumstances, that it cannot be calculated with any degree of accuracy. It has been estimated that a portion of the blood will make the tour of the body in about twenty-three seconds (*Flint*), and that the entire mass passes through the heart in from one to two minutes.†

Distribution and Regulation of the Heat of the Body.—1. DISTRIBUTION.—The natural temperature

---

* It is noticeable that the pulmonary set of veins circulates red blood, and the pulmonary set of arteries circulates dark blood. Both are connected with the lungs.

† If a salt which can be easily recognized be inserted in one of the jugular veins of a dog, and blood be drawn as quickly as possible from the opposite jugular, the substance will be detected in from twenty to thirty seconds ; having, in this brief time, passed to the right heart, thence to the lungs, back to the left heart, through the arteries, capillaries, and veins of the face and neck, and into the jugular vein. The total amount of blood in an adult of average weight is about eighteen pounds. Dividing this by five ounces, the quantity discharged by the left ventricle at each systole, gives fifty-eight pulsations as the number necessary to transmit all the blood in the body. This, however, is an extremely unreliable basis of calculation, as the rapidity of the blood is itself so variable.

is not far from 98°.* This is maintained, as we have already seen, by the action of the oxygen within us. Each capillary tube is a tiny stove, where oxygen is combining with the tissues of the body (see note, p. 105). Every contraction of a muscle develops heat, the latent heat being set free by the breaking up of the tissue. The warmth so produced is distributed by the circulation of the blood. Thus the arteries, veins, and capillaries form a series of hot-water pipes, through which the heated liquid is forced by a pump—the heart—while the heat is kept up, not by a central furnace and boiler, but by a multitude of little fires placed here and there along its course.

2. REGULATION.—The temperature of the body is regulated by means of the pores of the skin and the mucous membrane in the air-passages. When the system becomes too warm, the blood-vessels on the surface expand, the blood fills them, the fluid exudes into the perspiratory glands, pours out upon the exterior, and by evaporation cools the body.† When the temperature of the body is too low, the vessels contract, less blood goes to the surface, the perspiration decreases, and the loss of heat by evaporation diminishes.‡

---

* "The average temperature is, however, easily departed from. Through some trivial cause the cooling agencies may be interfered with, and then, the heating processes getting the superiority, a high temperature or fever comes on. Or the reverse may ensue. In Asiatic cholera, the constitution of the blood is so changed that its disks can no longer carry oxygen into the system, the heat-making processes are put a stop to, and, the temperature declining, the body becomes of a marble coldness characteristic of that terrible disease."—*Draper.*

† Just as water sprinkled on the floor cools a room. (*Physics*, page 191.)

‡ One can go into an oven where bread is baking, or into the arctic regions where the mountains are snow and the rivers ice, with equal impunity. Even by these extremes the temperature of the blood will be but slightly affected. In the one

Life by Death. — The body is being incessantly corroded, and portions borne away by the tireless oxygen. The scales of the epidermis are constantly falling off and being replaced by secretion from the cutis. The disks of the blood die, and new ones spring into being. On the continuance of this interchange depend our health and vigor. Every act is a destructive one. Not a bend of the finger, not a wink of the eye, not a thought of the brain but is at some expense of the machine itself. Every process of life is thus a process of death. The more rapidly this change goes on, and fresh, vigorous tissue takes the place of the old, the more elasticity and strength we possess.

Change of our Bodies.—There is a belief that our bodies change once in seven years. From the nature of the case, the rate must vary with the labor we perform; the organs most used altering oftenest. Probably the parts of the body in incessant employment are entirely reorganized many times within a single year.*

The Three Vital Organs. — Death is produced by the stoppage of the action of any one of the three organs—the heart, the lungs, or the brain. They have, therefore, been termed the "Tripod of Life." Really, however, as Huxley has remarked, "Life has but two legs to stand upon." If respiration

case, the flood-gates of perspiration will be opened and the superfluous heat expended in turning the water to vapor ; and, in the other, they will be tightly closed and all the heat retained.

* To use a homely simile, our bodies are like the Irishman's knife, which, after having had several new blades, and at least one new handle, was yet the same old knife.

and circulation be kept up artificially, the removal of the brain will not produce death.*

Wonders of the Heart.—The ancients thought the heart to be the seat of love. There were located the purity and goodness as well as the evil passions of the soul.† Modern science has found the seat of the mental powers to be in the brain. But, while it has thus robbed the heart of its romance, it has revealed wonders which eclipse all the mysteries of the past. This marvelous little engine throbs on continually at the rate of 100,000 beats per day, 40,000,000 per year, often 3,000,000,000 without a single stop. It is the most powerful of machines. "Its daily work is equal to one-third that of all the muscles. If it should expend its entire force in lifting its own weight vertically, it would rise 20,000 feet in an hour."‡ Its vitality is amazing. Lay upon a table the heart from a living sturgeon, all palpitating with life, and it will beat for days as if itself a living creature. The most tireless of organs while life exists, it is one of the last to yield when life expires. So long as a flutter lingers at the heart, we know the spark of being is not quite extinguished, and there is hope of restoration. During a life such as we sometimes see, it has propelled

---

* When death really does take place, i e., when the vital organs are stopped, it is noticeable that the tissues do not die for some time thereafter. If suitable stimulants be applied, as the galvanic battery, transfusion of blood, etc., the muscles may be made to contract, and many of the phenomena of life be exhibited.

† Our common words, hearty, large-hearted, courage (*cor*, the heart), are remains of this fanciful theory.

‡ "The greatest exploit ever accomplished by a locomotive, was to lift itself through less than one-eighth of that distance " Vast and constant as is this process, so perfect is the machinery that there are persons who do not even know where the heart lies until disease or accident reveals its location.

half a million tons of blood, yet repaired itself as
it has wasted, during its patient, unfaltering labor.
The play of its valves and the rhythm of its throb
have never failed until at the command of the
great Master-Workman the "wheels of life have
stood still."*

Fig. 43.

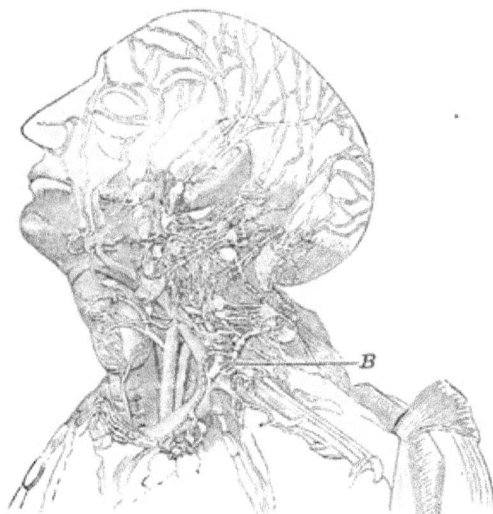

Lymphatics of the head and neck, showing the glands. and, B, the thoracic duct as it empties into the left innominate vein at the junction of the left jugular and subclavian veins.

The Lymphatic Circulation is intimately connected
with that of the blood. It is, however, more delicate
in its organization, and less thoroughly understood.
Nearly every part of the body is permeated by a

* "Our brains are seventy-five year clocks. The Angel of Life winds them up
once for all, then closes the case. and gives the key into the hand of the Angel of the
Resurrection. Tic-tac! tic-tac! go the wheels of thought ; our will cannot stop
them, they cannot stop themselves ; sleep cannot stop them ; madness only makes
them go faster ; death alone can break into the case, and, seizing the ever-swinging
pendulum which we call the heart. silence at last the clicking of the terrible escape-
ment we have carried so long beneath our wrinkled foreheads."—*Holmes.*

second series of capillaries, closely
interlaced with the blood-capilla-
ries already described, and
termed the Lymphatic system.
The larger number converge into
the thoracic duct—a small tube,
about the size of a goose-quill,
which empties into the great
veins of the neck (Fig. 43). Along
their course the lymphatics frequently pass through *glands,*—
hard, pinkish bodies of all sizes,
from that of a hemp-seed to an
almond. These glands are often
enlarged by disease, and then are
easily felt.

THE LYMPH, which circulates
through the lymphatics like blood
through the veins, is a thin,
colorless liquid, very like the
serum. This fluid, probably in
great measure an overflow from
the blood-vessels, is gathered up
by the lymphatics, undergoes in
the glands some process of preparation not well understood, and is
then returned to the circulation.

Office of the Lymphatics.—It is
thought that portions of the
waste matter of the body capable of further use are
thus, by a wise economy, retained and elaborated in
the system.

The *lacteals,* a class of lymphatics which will be

Fig. 44.

Lymphatics in the leg, with glands at the hip.

described under Digestion (p. 153), aid in taking
up the food ; after a meal they become milk-white.
In the lungs, the lymphatics are abundant ; sometimes absorbing the poison of disease, and diffusing
it through the system.*

The lymphatics of the skin we have already
spoken of as producing the phenomena of absorption.† Nature in her effort to heal a cut deposits an
excess of matter to fill up the breach. Soon, the
lymphatics go to work and remove the surplus material to other parts of the body.

Animals that hibernate are supported during the
winter by the fat which their absorbents carry into
the circulation from the extra supply they have laid
up during the summer. In famine or in sickness, a
man unconsciously consumes his own flesh.

Diseases, etc.—1. CONGESTION is an unnatural accumulation of blood in any part of the body. The
excess is indicated by the redness. If we put our
feet in hot water, the capillaries will expand by the
heat, and the blood set that way to fill them. The
red nose and purplish face of the drunkard show a
congestion of the capillaries. Those vessels have
lost their power of contraction, and so are permanently increased in size and filled with blood.
Blushing is a temporary congestion. The capillaries
being expanded only for an instant by the nervous
excitement, contract again and expel the blood.‡

---

* Persons have thus been poisoned by tiny particles of arsenic which evaporate
from green wall-paper, and float in the air.

† Pain is often relieved by infusing under the cuticle a solution of morphine,
which is taken up by the absorbents, and so carried through the system.

‡ " Blushing is a purely local modification of the circulation of this kind, and it will
be instructive to consider how a blush is brought about.   An emotion—sometimes

2. INFLAMMATION means simply a burning. If
there is irritation or an injury at any spot, the blood
sets thither and reddens it. This extra supply, both
by its presence and the friction of the swiftly-moving
currents, produces heat. The pressure of the distended vessels upon the nerves frets them, and produces pain. The swelling stretches the walls of the
blood-vessels, and the serum or lymph oozes through.
The four characteristics of an inflammation are redness, heat, pain, and swelling.

3. BLEEDING, if from an artery, will be of red
blood, and will come in jets ; if from the veins, it
will be of dark blood, and will flow in a steady
stream. If only a small vessel be severed, it may be
checked by a piece of cloth held or bound firmly

pleasurable, sometimes painful—takes possession of the mind ; thereupon a hot flush
is felt, the skin grows red, and according to the intensity of the emotion these changes
are confined to the cheeks only, or extend to the ' roots of the hair,' or ' all over.' What
is the cause of these changes? The blood is a red and a hot fluid ; the skin reddens and
grows hot, because its vessels contain an increased quantity of this red and hot fluid :
and its vessels contain more, because the small arteries suddenly dilate. the natural
moderate contraction of their muscles being superseded by a state of relaxation. In
other words, the action of the nerves which cause this muscular contraction is suspended. On the other hand, in many people, extreme terror causes the skin to grow
cold, and the face to appear pale and pinched. Under these circumstances, in fact,
the supply of blood to the skin is greatly diminished, in consequence of an excessive
stimulation of the nerves of the small arteries, which causes them to contract and so
to cut off the supply of blood more or less completely. That this is the real state of
the case may be proved experimentally upon rabbits. These animals, it is true, do
not blush naturally, but they may be made to blush artificially. If, in a rabbit, the
sympathetic nerve which sends branches to the vessels of the head is cut. the ear of
the rabbit, which is covered by so delicate an integument that the changes in its
vessels can be readily perceived, at once blushes. That is to say, the vessels dilate,
fill with blood, and the ear becomes red and hot. The reason of this is, that when
the sympathetic nerve is cut, the nervous stimulus which is ordinarily sent along its
branches is interrupted, and the muscles of the small vessels, which were slightly
contracted, become altogether relaxed. And now it is quite possible to produce
pallor and cold in the rabbit's ear. To do this it is only necessary to irritate the cut
end of the sympathetic nerve which remains connected with the vessels. The nerve
then becomes excited. so that the muscular fibers of the vessels are thrown into a
violent state of contraction, which diminishes their caliber so much that the blood
can hardly make its way through them. Consequently, the ear becomes pale and
cold."—*Huxley's Lessons in Physiology*, page 58.

upon the wound. If a large trunk be cut, especially in a limb, make a knot in a handkerchief and tie it loosely about the limb; then, placing the knot on the wound, with a short stick twist the handkerchief tightly enough to stop the flow. If you have a piece of cloth to use as a pad, the knot will be unnecessary. If it be an artery that is cut, the pressure should be applied between the wound and the heart; if a vein, beyond the wound. If you are alone, and are severely wounded, or in an emergency, like a railroad accident, use the remedy which has saved many a life upon the battle-field—bind or hold a handful of dry earth upon the wound, elevate the part, and await surgical assistance.

4. SCROFULA is generally inherited. It is a disease affecting the lymphatic glands, most commonly those of the neck, forming " kernels," as they are called. It is, however. liable to attack any organ, and frequently terminates in consumption. Persons inheriting this disease can hope to ward off its insidious approaches only by the utmost care in diet and exercise; by the use of pure air and warm clothing. and by avoiding late hours and undue stimulus of all kinds. Probably the most fatal and common excitants of the latent seeds of scrofula are insufficient or improper food, and want of ventilation.

5. A COLD.—We put on a thinner dress than usual, or, when heated, sit in a cool place. The skin is chilled, and the perspiration checked. The blood, no longer cleansed, and reduced in volume by the drainage through the pores, sets to the lungs for purification. That organ is oppressed, breathing becomes difficult, and the extra mucous secreted by

the irritated surface of the membrane is thrown off
by coughing. The mucous membrane of the nasal
chamber sympathizes with the difficulty, and we
have "a cold in the head," or a catarrh. In
general, the excess of blood seeks the weakest point,
and develops there any latent disease.* Where one
person has been killed in battle, thousands have
died of colds.

To restore the equipoise must be the object of all
treatment. We put the feet in hot water and they
soon become red and gorged with the blood which is
thus called from the congested organs. Hot foot-
baths have saved multitudes of lives. It is well in
case of a sudden cold to go immediately to bed, and
with hot drinks and extra clothing open the pores,
and induce free perspiration. This calls the blood to
the surface, and, by equalizing and diminishing the
volume of the circulation, affords relief.†

The rule for the prevention and cure of a cold is to
keep the blood upon the surface.

6. CATARRH commonly manifests itself by the
symptoms known as those of a "cold in the head,"
and is produced by the same causes. It is an inflammation of the mucous membrane lining the
nasal and bronchial passages. One going out from

---

* A party go out for a walk and are caught in a rain, or, coming home heated from
some close assembly, throw off their coats to enjoy the deliciously-cool breeze. The
next day, one has a fever, another a slight headache, another pleurisy, another
pneumonia, another rheumatism, while some escape without any ill-feeling whatever. The last had vital force sufficient to withstand the disturbance, but in the
others there were weak points, and to these the excess of blood has gone, producing
congestion.

† Severe colds may often be relieved in their first stages by using lemons freely
during the day, and taking at night fifteen or twenty grains of sodium bromide.
Great care, however, should be observed in employing the latter remedy, except
under the advice of a physician.

the hot dry air of a furnace-heated room into the cold damp atmosphere of our climate can hardly avoid irritating and inflaming this tender membrane. If our rooms were heated less intensely, and ventilated more thoroughly, so that we had not the present hot-house sensitiveness to cold air, this disease would be far less universal, and perhaps would disappear entirely.*

ALCOHOLIC DRINKS AND NARCOTICS.

1. ALCOHOL.†

General Effect of Alcohol upon the Circulation. —During the experiment described on page 116, the influence of alcohol upon the blood may be beautifully tested. Place on the web of the frog's

* Dr. Gray gives the following table based upon measurement of rooms occupied by letter-press printers :

	Number per cent. Spitting Blood.	Subject to Catarrh.
104 men having less than 500 cubic feet of air to breathe...........	12.50	12.50
115 men having from 500 to 600 cubic feet of air to breathe.	4.35	3 58
101 men having more than 600 cubic feet of air to breathe	3.96	1.99

† How Alcohol is formed.—When any substance containing sugar, as fruit-juice, is caused to ferment, the elements of hydrogen. carbon, and oxygen, of which the sugar is composed, rearrange themselves so as to form carbonic acid, alcohol, and certain volatile oils. and ethers. The carbonic acid partly evaporates, and partly remains to give life and piquancy to the liquor ; the alcohol is the exciting or intoxicating principle; while the oils and ethers impart the peculiar flavor and aroma. Thus wine is fermented grape-juice and cider is fermented apple-juice. each having its distinctive fragrance. For a full account of the subject of Fermentation, read *Steele's New Chemistry*, page 192.

Manufacture of Beer —The barley used for making beer is first *malted*. i. e. sprouted, to turn a part of its starch into sugar. When this process has gone far foot a drop of dilute spirit. The blood-vessels immediately expand—an effect known as "*Vascular enlargement.*" Channels before unseen open, and the blood-disks fly along at a brisker rate. Next, touch the membrane with a drop of pure spirit. The blood channels quickly contract; the cells slacken their speed; and, finally, all motion ceases. The flesh shrivels up and dies. The circulation thus stopped is stopped forever. The part affected will in time slough off. Alcohol has killed it.

enough, it is checked by heating the grain in a kiln until the germ is destroyed. The malt is then crushed, steeped, and fermented with hops and yeast. The sugar gradually disappears, alcohol is formed, and carbonic acid escapes into the air. The beer is then put into casks, where it undergoes a second, slower fermentation, the flavor ripens, and the carbonic acid gathers; when the liquor is drawn, this gas bubbles to the surface, giving to the beer its sparkling, foamy look.

Spirits.—Alcohol is so volatile that, by the application of heat, it can be driven off as a vapor from the fermented liquid in which it has been produced. Steam and various fragrant substances will pass over with it, and, if they are collected and condensed in a cool receiver, a new and stronger liquor will be formed, having a distinctive odor.

In this way, the alcohol of commerce is distilled from whisky; brandy, from wine; rum, from fermented molasses; whiskey, from fermented corn, barley or potatoes; and gin, from fermented barley and rye, afterward distilled with juniper berries. In all liquors, the base is alcohol. It comprises from 3 to 8 per cent. of ale and porter, 7 to 17 per cent. of wine, and 40 to 50 per cent. of brandy and whisky. They may therefore be considered as alcohol more and less diluted with water and flavored with various aromatics. The taste, agreeability, etc., of different liquors—as brandy, gin, beer, cider, etc.—may vary greatly, but they all produce certain physiological effects due to their common ingredient—alcohol.

Properties of Alcohol.—Pour a little alcohol into a saucer and apply an ignited match. The liquid will suddenly take fire, burning with intense heat, but feeble light. In this process, alcohol takes up oxygen from the air, forming carbonic acid gas, and water.—Hold a red-hot coil of platinum wire in a goblet containing a few drops of alcohol, and a peculiar odor will be noticed. It denotes the formation of *aldehyde*—a substance produced in the slow oxidation of alcohol. Still further oxidized, the alcohol would be changed into *acetic acid*—the sour principle of vinegar.

One of the most noticeable properties of alcohol is its affinity for water. When strong alcohol is exposed to the air, it absorbs moisture and becomes diluted; at the same time, the spirit itself evaporates. The commercial or proof-spirit is about one-half water; the strongest holds ten per cent.; and to obtain absolute or waterless alcohol, requires careful distillation in connection with some substance, as lime, that has a still greater affinity for water, and so can despoil the alcohol.—Put the white of an egg—nearly pure albumen—into a cup, and pour upon it some alcohol, or even strong brandy; the fluid albumen will coagulate, becoming hard and solid. In this connection, it is well to remember that albumen is contained in our food, while the brain is largely an albuminous substance.

The influence of alcohol upon the human system is similar. Alcohol is a poison. A quart drunk at a time, would kill a man like a bullet. Diluted, as in wine or whiskey, it dilates the blood-vessels, quickens the circulation, hastens the heart-throbs, and accelerates the respiration.

The Effect of Alcohol upon the Heart. — What means this rapid flow of the blood? It shows that the heart is overworking. The nerves that lead to the minute capillaries and regulate the passage of the vital current through the extreme parts of the body, are paralyzed by this active narcotic. The tiny blood-vessels at once expand. This "Vascular enlargement" removes the resistance to the passage of the blood, and hence to the beat of the heart, and the heart flies like the main spring of a clock when the wheels are taken out.*

Careful experiments show that two ounces of alcohol—an amount contained in the daily potations of a very moderate ale or whisky drinker—increase the heart-beats 6000 in twenty-four hours ;—a degree of work represented by that of lifting up a weight of seven tons to a height of one foot. Reducing this sum to ounces and dividing, we find that the heart is

---

* In the text, the researches of Dr. B. W. Richardson have been accepted as authoritative on both sides of the Atlantic. It should, however, be noted that Dr. Palmer, of the University of Michigan, claims that alcohol does not, at any time, increase, but, instead, diminishes the action of the heart. Prof. Martin, of Johns Hopkins University, from a series of experiments upon dogs, concludes that "blood containing one eighth per cent. by volume of absolute alcohol has no immediate action on the isolated heart. Blood containing one-fourth per cent. by volume of absolute alcohol, almost invariably remarkably diminishes within a minute the work done by the heart ; blood containing one-half per cent. always diminishes it, and even may bring the amount pumped out by the left ventricle to so small a quantity that it is not sufficient to supply the coronary arteries ; hence blood is drained off by them from the outflow tube and at last none is pumped out from its upper end at all." (See note, p. 193.)

driven to do extra work equivalent to lifting seven ounces one foot high 1493 times each hour! No wonder that the drinker feels a reaction, a physical languor, after the earliest effects of his indulgence have passed away. The heart flags, the brain and the muscles feel exhausted, and rest and sleep are imperatively demanded. During this time of excitement, the machinery of life has really been "running down." "It is hard work," says Richardson, " to fight against alcohol; harder than rowing, walking, wrestling, coal-heaving, or the tread-mill itself."

All this is only the first effect of alcohol upon the heart. Long-continued use of this disturbing agent causes a " Degeneration of the muscular fiber,"* so that the heart loses its old power to drive the blood, and, after a time, fails to respond even to the spur of the excitant that has urged it to ruin.

Influence upon the Membranes.—The flush of the face and the blood-shot eye, that are such noticeable effects of even a small quantity of liquor, indicate the condition of all the internal organs. The delicate linings of the stomach, heart, brain, liver, and lungs, are reddened, and every tiny vein is inflamed, like the blushing nose itself. If the use of liquor is

---

* This " Degeneration " of the various tissues of the body, we shall find, as we proceed. is one of the most marked effects of alcoholized blood. The change consists in an excess of liquid, or, more commonly, in a deposit of fat. This fatty matter is not an increase of the organ, but it takes the place of a part of its fiber, thus weakening the structure, and reducing the power of the tissue to perform its function. Almost everywhere in the body we thus find cells—muscle-cells, liver-cells, nerve-cells, as the case may be—changing, one by one, under the influence of this potent disorganizer, into unhealthy fat-cells. " Alcohol has well been termed," says the London Lancet, "the ' Genius of Degeneration.' "

The cause of this degeneration can be easily explained. The increased activity of the circulation compels a correspondingly-increased activity of the cell-changes: but the essential condition of healthful change—the presence of additional oxygen—is wanting (see p. 133), and the operation is imperfectly performed.—(*Brodie.*)

habitual, this "Vascular enlargement," that at first slowly passed away after each indulgence, becomes permanent, and now the discolored, blotched skin reveals the state of the entire mucous membrane.

We learned on page 55 what a peculiar office the membrane fills in nourishing the organs it enwraps. Anything that disturbs its delicate structure must mar its efficiency. Alcohol has a wonderful affinity for water. To satisfy this greed, it will absorb moisture from the tissues with which it comes in contact, as well as from their lubricating juices. The enlargement of the blood-vessels and their permanent congestion must interfere with the filtering action of the membrane. In time, all the membranes become dry, thickened, and hardened; they then shrink upon the sensitive nerve, or stiffen the joint, or enfeeble the muscle. The function of these membranes being deranged. they will not furnish the organs with perfected material, and the clogged pores will no longer filter their natural fluids. Every organ in the body will feel this change.

Effect upon the Blood.*—From the stomach, alcohol passes directly into the circulation, and so, in a few minutes, is swept through the entire system. If it be present in sufficient amount and strength, its eager desire for water will lead it to absorb moisture from the red corpuscles, causing them to shrink. change their form, harden. and lose some of their ability to carry oxygen : it may even

---

* Dr. G. B. Harriman of Boston states, as the result of his observations, that alcohol acts upon the oxygen-carrier, the coloring matter of the red corpuscles, causing it to settle in one part of the globule, or even to leave the corpuscle, and deposit itself in other elements of the blood. Thus the red corpuscle may become colorless, distorted, shrunken, and even entirely broken up.

make them adhere in masses, and so hinder their passage through the tiny capillaries.—*Richardson.*

With most persons who indulge freely in alcoholic drinks, the blood is thin, the avidity of alcohol for water causing a burning thirst so familiar to all drinkers, and hence the use of enormous quantities of water, oftener of beer, which unnaturally dilutes the blood. The blood then easily flows from a wound, and, not coagulating, renders an accident or surgical operation very dangerous.

When the blood tends, as in the case of an excessive use of spirits, to coagulate in the capillaries, there is a liability of an obstruction to the flow of the vital current through the heart,* liver, lungs, etc., that may cause disease, and in the brain may lay the foundation of paralysis, or, in extreme cases, of apoplexy.

Wherever the alcoholized blood goes through the body, it bathes the delicate cells with an irritating, narcotic poison, instead of a bland, nutritious substance.

Effect upon the Lungs. — Here we can see how certainly the presence of alcohol interferes with the red corpuscles in their task of carrying oxygen. "Even so small a quantity as one part of alcohol to 500 of the blood will materially check the absorption of oxygen in the lungs."

The cells, unable to take up oxygen, retain their carbonic acid gas, and so return from the lungs, carrying back, to poison the system, the refuse matter

---

* Persons have drunk a large quantity of liquor for a wager, and, as the result of their folly, "died upon the spot." The whole of the blood in the heart being turned into a clot, the circulation was instantly stopped, and death was instantaneous.

the body has sought to throw off. Thus the lungs no longer furnish properly oxygenized blood.

The rapid stroke of the heart, already spoken of, is followed by a corresponding quickening of the respiration. The flush of the cheek is repeated in the reddened mucous membrane lining the lungs.

When this "Vascular enlargement" becomes permanent, and the highly-albuminous membrane of the air-cells is hardened and thickened as well as congested, the Osmose of the gases to and fro through its pores can no longer be prompt and free as before. Even when the effect passes off in a few days after the occasional indulgence, there has been, during that time, a diminished supply of the life-giving oxygen furnished to the system ; weakness follows, and, in the case of hard drinkers, there is a marked liability to epidemics.*

Physicians tell us, also, that there is a peculiar form of consumption known as Alcoholic Phthisis caused by long-continued and excessive use of liquor. It generally attacks those whose splendid physique has enabled them to "drink deep" with apparent impunity. This type of consumption appears late in life and is considered incurable. Severe cases of pneumonia are also generally fatal with inebriates.

* "There is no doubt that alcohol alters and impairs tissues so that they are more prone to disease."—(Dr. G. K. Sabine.) A volume of statistics could be filled with quotations like the following : "Mr. Huber, who saw in one town in Russia two thousand one hundred and sixty persons perish with the cholera in twenty days, said : 'It is a most remarkable circumstance that persons given to drink have been swept away like flies. In Tiflis, with twenty thousand inhabitants, every drunkard has fallen,—all are dead, not one remaining.'"

† The Influence of Alcohol is continued in the chapter on Digestion.

PRACTICAL QUESTIONS.

1. Why does a dry, cold atmosphere favorably affect catarrh?

2. Why should we put on extra covering when we lie down to sleep?

3. Is it well to throw off our coats or shawls when we come in heated from a long walk?

4. Why are close-fitting collars or neck-ties injurious?

5. Which side of the heart is the more liable to inflammation?

6. What gives the toper his red nose?

7. Why does not the arm die when the surgeon ties the principal artery leading to it?

8. When a fowl is angry, why does its comb redden?

9. Why does a fat man endure cold better than a lean one?

10. Why does one become thin during a long sickness?

11. What would you do if you should come home "wet to the skin"?

12. When the cold air strikes the face, why does it first blanch and then flush?

13. What must be the effect of tight lacing upon the circulation of the blood?

14. Do you know the position of the large arteries in the limbs, so that in case of accident you could stop the flow of blood?

15. When a person is said to be good-hearted, is it a physical truth?

16. Why does a hot foot-bath relieve the headache?

17. Why does the body of a drowned or strangled person turn blue?

18. What are the little "kernels" in the arm-pits?

19. When we are excessively warm, would the thermometer show any rise of temperature in the body?

20. What forces besides that of the heart aid in propelling the blood?

21. Why can the pulse be best felt in the wrist?

22. Why are starving people exceedingly sensitive to any jar?

23. Why will friction, an application of horse-radish leaves, or a blister, relieve internal congestion?

24. Why are students very liable to cold feet?

25. Is the proverb that "blood is thicker than water" literally true?

26. What is the effect upon the circulation of "holding the breath"?

27. Which side of the heart is the stronger?

28. How is the heart itself nourished ?*

29. Does any venous blood reach the heart without coming through the venæ cavæ ?

30. What would you do, in the absence of a surgeon, in the case of a severe wound ? See Appendix.

31. What would you do in the case of a fever ? See Appendix.

32. What is the most injurious effect of alcohol upon the blood ?

33. Are our bodies the same from day to day ?

34. Show how life comes by death.

35. Is not the truth just stated as applicable to moral and intellectual, as to physical life ?

36. What vein begins and ends with capillaries? *Ans.* The portal vein commences with capillaries in the digestive organs, and ends with the same kind of vessels in the liver. (See p. 153.)

---

* The coronary artery, springing from the aorta just after its origin, carries blood to the muscular walls of the heart ; the venous blood comes back through the coronary veins, and empties directly into the right auricle.

VI.

DIGESTION

AND

FOOD.

A man puts some ashes in a hill of corn and thereby doubles its yield. Then he says, " My ashes have I turned into corn." Weak from his labor, he eats of his corn, and new life comes to him. Again, he says, " I have changed my corn into a man." This also he feels to be the truth.

It is the problem of the body, remember, that we are discussing. A man is more than the body ; to confound the body and the man is worse than confounding the body and the clothing.—JOHN DARBY.

BLACKBOARD ANALYSIS.

DIGESTION AND FOOD.

1. WHY WE NEED FOOD.

2. WHAT FOOD DOES.

3. KINDS OF FOOD. . ...
- 1. Nitrogenous.
- 2. Carbonaceous......
  - a. The Sugars.
  - b. The Fats.
- 3. Minerals.

4. ONE KIND IS INSUFFICIENT.

5. OBJECT OF DIGESTION.

6. PROCESS OF DIGESTION
- — General Description.
- 1. Mastication and Insalivation.......
  - a. The Saliva.
  - b. Process of Swallowing.
- 2. Gastric Digestion..
  - a. The Stomach.
  - b. The Gastric Juice.
  - c. The Chyme.
- 3. Intestinal Digestion.
  - — Description.
  - a. The Bile.
  - b. The Pancreatic Juice.
  - c. The Small Intestine.
- 4. Absorption........
  - a. By the Veins.
  - b. By the Lacteals.

7. COMPLEXITY OF THE PROCESS OF DIGESTION.

8. HYGIENE ....
- 1. Length of Time required.
- 2. Value of different kinds of Food.
  - a. Beef.
  - b. Mutton.
  - c. Lamb.
  - d. Pork.
  - e. Fish.
  - f. Milk.
  - g. Cheese.
  - h. Eggs, etc.
- 3. The Stimulants.
  - a. Coffee.
  - b. Tea.
  - c. Chocolate.
- 4. Cooking of Food.
- 5. Rapid Eating.
- 6. Quantity and Quality of Food.
- 7. When Food should be taken.
- 8. How " " "
- 9. Need of a Variety.

9. THE WONDERS OF DIGESTION.

10. DISEASES............
- 1. Dyspepsia.
- 2. The Mumps.

11. ALCOHOLIC DRINKS AND NARCOTICS.
1. Alcohol.
- 1. Is Alcohol a Food?
- 2. Effect upon the Digestion.
- 3. " " " Liver.
- 4. Does Alcohol impart heat?
- 5. " " " strength?
- 6. The Effect upon the Waste of the Body.
- 7. Alcohol creates a progressive appetite for itself.

DIGESTION AND FOOD.

WHY **we need Food.**—We have learned that our bodies are constantly giving off waste matter—the products of the fire, or oxidation, as the chemist terms the change going on within us (note, p. 105). A man without food will starve to death in a few days, i. e., the oxygen will have consumed all the available flesh of his body.* To replace the daily outgo, we need about two and a quarter pounds of food, and three pints of drink.†

Including the eight hundred pounds of oxygen taken from the air, a man uses in a year about a ton

---

* The stories current in the newspapers of persons who live for years without food, are, of course, untrue. The case of the Welsh Fasting Girl, which excited general interest throughout Great Britain, and was extensively copied in our own press, is in point. She had succeeded in deceiving not only the public but, as some claim, her own parents. At last a strict watch was set by day and night, precluding the possibility of her receiving any food except at the hands of the committee, from whom she steadily refused it. In a few days she died from actual starvation. The youth of the girl, the apparent honesty of the parents, and the tragical sequel, make it one of the most remarkable cases of the kind on record.

† "From experiments performed while living on an exclusive diet of bread, fish, meat, and butter, with coffee and water for drink, we have found that the entire quantity of food required during twenty-four hours, by a man in full health, and taking free exercise in the open air, is as follows :

Meat	16 ounces or 1.00 lb. avoirdupois.	
Bread	19 " " 1.19 lbs. "
Butter, or fat . .	3½ " " 0.22 lbs. "
Water	52 fluid oz " 3.38 lbs. "

That is to say, rather less than two and a half pounds of solid food, and rather over three pints of liquid food." –*Dalton.*

and a half of material.* Yet during this entire time his weight may have been nearly uniform.† Our bodies are but molds, in which a certain quantity of matter, checked for a time on its ceaseless round, receives a definite form. They may be likened, says Huxley, to an eddy in the river, which retains its shape for a while, yet every instant each particle of water is changing.

What Food Does.—We make no force ourselves. We can only use that which nature provides.‡ All our strength comes from the food we eat. Food is force—that is, it contains latent within it a power which it gives up when it is decomposed.§ Oxygen is the magic key which unlocks for our use this hid-

---

* The following is the daily ration of a United States soldier. It is said to be the most generous in the world :—

Bread or flour	22 ounces.
Fresh or salt beef (or pork or bacon, 12 oz.)	20 "
Potatoes (three times per week)	16 "
Rice	1.6 "
Coffee (or tea, 0.24 oz.)	1.6 "
Sugar	2.4 "
Beans	0.64 gill.
Vinegar	0.32 "
Salt	0.16 "

† If, however, he were kept on the scale-pan of a sensitive balance, he would find that his weight is constantly changing, increasing with each meal, and then gradually decreasing.

‡ We draw from nature at once our substance, and the force by which we operate upon her; being, so far, parts of her great system, immersed in it for a short time and to a small extent. Enfolding us, as it were, within her arms, Nature lends us her forces to expend; we receive them, and pass them on, giving them the impress of our will, and bending them to our designs, for a little while ; and then—— Yes ; then it is all one. The great procession pauses not, nor flags a moment, for our fall. The powers which Nature lent to us she resumes to herself, or lends, it may be, to another ; the use which we have made of them, or might have made and did not, written in her book for ever.—*Health and its Conditions.*

§ This force is chemical affinity. It binds together the molecules which compose the food we eat. When oxygen tears the molecules to pieces and makes them up into smaller ones the force is set free. As we shall learn in Physics, it can be turned into heat, muscular motion, electricity, etc. The principle that the different kinds of force can be changed into one another without loss, is called the Conservation of energy, and is one of the grandest discoveries of modern science. (Physics, pages 37, 40, 208.)

den store.*   Putting food into our bodies is like
placing a tense spring within a watch ; every motion
of the body is only a new direction given to this
food-force, as every movement of the hand on the
dial is but the manifestation of the power of the
bent spring in the watch.   We use the pent-up energies of meat, bread, and vegetables which are placed
at our service, and transfer them to a higher theater
of action.†

Kinds of Food Needed. — From what has been
said it is clear that, in order to produce heat and
force, we need something that will burn, i. e., with
which oxygen can combine.   Experiment has
proved that to build up every organ, and keep the
body in the best condition, we require three kinds
of food.

1. NITROGENOUS FOOD, or that which contains
much nitrogen, is a prominent constituent of the
tissues of the body. and is therefore necessary to
their growth and repair.‡   The most common forms
are whites of eggs—which are nearly pure albumen ;
casein—the chief constituent of cheese ; lean meat ;
and gluten—the viscid substance which gives tenac-

---

* We have spoken of the mystery that envelops the process of the conversion
of food-force into muscular-force (note, p. 108). All physiologists agree that muscular power has its source in the chemical decomposition of certain substances
whereby their potential energy is released. Probably some of the food undergoes
this chemical change before it passes out of the alimentary canal ; possibly some is
broken up by the oxygen while it is being swept along by the blood ; but, probably
by far the largest part is converted into the various tissues of the body. and finally
becomes a waste product only after there takes place in the tissue itself that chemical disorganization that sets free its stored-up power.—*Foster's Physiology*.

† It is a grand thought that we can thus transform what is common and gross
into the refined and spiritual ; that out of waving wheat, wasting flesh. running
water, and dead minerals, we can realize the glorious possibilities of human life.

‡ Since this kind of food closely resembles albumen, it is sometimes called
Albuminous. The term Proteid is also used.

ity to dough. Bodies having a great deal of nitrogen readily oxidize. Hence the peculiar character of the quick-changing, force-exciting muscle.

2. CARBONACEOUS FOOD—i. e., food containing much carbon, consists of two kinds—viz., the *sugars*, and the *fats*.

(1.) *The sugars* contain hydrogen and oxygen in the proportion to form water, and about the same amount of carbon. They may, therefore, be considered as water, with carbon diffused through it. In digestion, starch and gum are changed to sugar, and so are ranked with this class.

(2) *The fats* are like the sugars in composition, but contain less oxygen, and not in the proportion to form water. They combine with more oxygen in burning, and so give off more heat.

The non-nitrogenous elements of the food have, however, other uses than to develop heat.* Fat is essential to the assimilation of the food, while sugar and starch aid in digestion and may be converted into fat.† Fat and carbonaceous material both enter into the composition of the various tissues, and when, by the breaking-up of the contractile substance of the muscle, their latent energy is set free. they become the source of muscular force, as well as heat. While the tendency of the albuminous food is to excite chemical action, and hence the release of energy, the fats and carbonaceous food

---

* The heat they produce in burning may be turned into motion of the muscles, according to the principle of the Conservation of energy (p. 140, note) ; while all the structures of the body in their oxidation develop heat.

† In Turkey, the ladies of the harem are fed on honey and thick gruel, to make flesh, which is considered to enhance their beauty.—The negroes on the sugar plantations of the South always grow fat during the sugar-making season.

may be laid up in the body to serve as a storehouse of energy to supply future needs.

3. MINERAL MATTERS.—Food should contain water, and certain common minerals, such as iron,* sulphur, magnesia, phosphorus, salt, and potash. About three pints of water are needed daily to dissolve the food and carry it through the circulation, to float off waste matter, to lubricate the tissues, and by evaporation to cool the system. It also enters largely into the composition of the body. A man weighing 154 pounds contains 100 pounds of water, about 12 gallons,—enough, if rightly arranged, to drown him.†

Iron goes to the blood disks; lime combines with phosphoric and carbonic acids to give solidity to the bones and teeth; phosphorus is essential to the activity of the brain. Salt is necessary to the secretion of some of the digestive fluids, and also to aid in working off from the system its waste products. These various minerals, except iron—sometimes given as a medicine, and salt—universally used as a condiment,‡ are contained in small, but

---

* While the body can build up a solid from liquid materials on the one hand, on the other it can pour iron through its veins and reduce the hardest textures to blood. —*Hinton*.

† It is said that Blumenbach had a perfect mummy of an adult Teneriffian, which with the viscera weighed only seven and a half pounds.

‡ "Animals will travel long distances to obtain salt. Men will barter gold for it; indeed, among the Gallas and on the coast of Sierra Leone, brothers will sell their sisters, husbands their wives, and parents their children for salt. In the district of Accra, on the gold coast of Africa, a handful of salt is the most valuable thing upon earth after gold, and will purchase a slave. Mungo Park tells us that with the Mandingoes and Bambaras the use of salt is such a luxury that to say of a man 'he flavors his food with salt,' it is to imply that he is rich; and children will suck a piece of rock-salt as if it were sugar. No stronger mark of respect or affection can be shown in Muscovy, than the sending of salt from the tables of the rich to their poorer friends. In the book of Leviticus it is expressly commanded as one

sufficient quantities in meat, bread, and vegetables.

One Kind of Food is Insufficient.—A person fed
on starch alone, would die. It would be a clear case
of nitrogen starvation. On the other hand, as nitrogenous food contains carbon, the elements of water,
and various mineral matters, life could be supported
on that alone. But such a prodigious quantity of
lean meat, for example, would be required to furnish
the other elements, that not only would it be very
expensive, but it is likely that after a time the labor
of digestion would be too onerous, and the system
would give up the task in despair. The need of a
diet containing both nitrogenous and carbonaceous
elements is shown in the fact that even in the trop-
ical regions oil is relished as a dressing upon salad.
Instinct everywhere suggests the blending. Butter
is used with bread ; rice is boiled with milk ; cheese‘
is eaten with maccaroni, and beans are baked with
pork.

The Object of Digestion.—If our food were cast
directly into the blood, it could not be used. For
example, although the chemist cannot see wherein
the albumen of the egg differs from the albumen of
the blood, yet if it be injected into the veins it is

---

of the ordinances of Moses, that every oblation of meat upon the altar shall be
seasoned with salt, without lacking ; and hence it is called the Salt of the Covenant
of God. The Greeks and Romans also used salt in their sacrificial cakes ; and it is
still used in the services of the Latin church—the '*parra mica*,' or pinch of salt,
being in the ceremony of baptism, put into the child's mouth, while the priest says,
'Receive the salt of wisdom, and may it be a propitiation to thee for eternal life.'
Everywhere and almost always, indeed, it has been regarded as emblematical of
wisdom, wit, and immortality. To taste a man's salt, was to be bound by the rites
of hospitality ; and no oath was more solemn than that which was sworn upon
bread and salt. To sprinkle the meat with salt was to drive away the devil, and to
this day, nothing is more unlucky than to spill the salt."—*Letheby, On Food.*

unavailable for the pur-
poses required, and is
thrown out again. In the
course of digestion the
food is modified in various
ways whereby it is fitted
for the use of the body.
We call this process *assimilation*—a name for a work
done solely by the vital
organs and so mysterious
in its nature that the
wisest physiologist gets
only glimpses here and
there of its operations.

**The General Plan of Di-
gestion.** — Nature has provided for this purpose an
entire laboratory, furnished
with a chemist's outfit of
knives, mortars, baths,
chemicals, filters, etc. The
food is (1) chewed, mixed

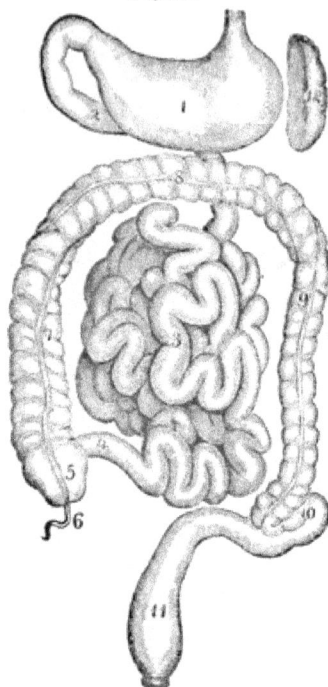

Fig. 45.

The *Stomach and Intestines.* 1, *stomach ;* 2, *duodenum ;* 3, *small intestine ;* 4, *termination of the ileum :* 5, *cæcum ;* 6, *vermiform appendix ;* 7, *ascending colon ;* 8, *transverse colon ;* 9, *descending colon ;* 10, *sigmoid flexure of the colon ;* 11, *rectum ;* 12, *spleen—a gland whose action is not understood.*

with the saliva in the mouth, and swallowed ; (2) it is
acted upon by the gastric juice in the stomach ; (3)
passed into the intestines, where it receives the bile,
pancreatic juice, and other liquids which completely
dissolve it ;* (4) the nourishing part is absorbed in
the stomach and intestines, and thence thrown into
the blood-vessels, whence it is whirled through the

* "Digestion." says Berzelius, " is a process of rinsing." The digestive apparatus
secretes, and again absorbs with the food which it has dissolved, not less than three
gallons of liquid per day.—*Barnard, Bidder, Schmidt,* and others.

Fig. 46.

The Parotid—one of the salivary glands.

body by the torrent of the circulation. These processes take place within the *alimentary canal*, a narrow, tortuous tube which commences at the mouth, and is about thirty feet long.*

I. Mastication and Insalivation.—1. THE SALIVA.— The food while being cut and ground by the teeth is mixed with the saliva. This is a thin, colorless, frothy, slightly alkaline liquid, secreted† by the mucous membrane lining the mouth, and by three pairs of salivary glands (parotid, submaxillary, and sublingual) opening into the mouth through ducts, or tubes. The amount varies, but on the average is about three pounds per day, and in health is always

---

* The digestive apparatus is lined with mucous membrane, that possesses functions similar to those of the outer skin. It absorbs certain substances and rejects waste matter. On account of this close connection between the inner and the outer skin, it is not surprising to find that in the lowest animals digestion is performed by means of the external skin. The amœba, which is merely a gelatinous mass, when it takes its food, extemporizes a stomach for the occasion. It simply wraps itself around the morsel, and, like an animated apple-dumpling with the apple for food and the crust for animal, goes on with the process until the operation is completed, when it unrolls itself again and lets the indigestible residue escape. The common hydra of our brooks can live when turned inside out, like a glove: either side serving for skin or stomach, as necessity requires.

† By secretion is meant merely a separation or picking out from the blood.

sufficient to keep the mouth moist.* It softens and dissolves the food, and thus enables us to get the flavor or taste of what we eat. It contains a peculiar organic principle called *ptyalin*,† which, acting upon the starch of the food, changes it into glocose or grape-sugar.

2. THE PROCESS OF SWALLOWING.—The food thus finely pulverized, softened, and so lubricated by the viscid saliva as to prevent friction as it passes over the delicate membranes, is conveyed by the tongue and cheek to the back of the mouth. The soft palate lifts to close the nasal opening ; the epiglottis shuts down, and along this bridge the food is borne, without danger of falling into the windpipe or escaping into the nose. The muscular bands of the throat now seize it and take it beyond our control. The fibers of the œsophagus contract above, while they are lax below, and convey the food by a worm-like motion into the stomach.‡

II. Gastric Digestion.—1. THE STOMACH is an irregular expansion of the digestive tube. Its shape has

---

* The presence and often the thought of food will " make one's mouth water." Fear checks the flow of saliva, and hence the East Indians sometimes attempt to detect theft by making those who are suspected chew rice. The person from whom it comes out driest is adjudged the thief !

† One part of ptyalin will convert 8,000 parts of starch into sugar.—*Mialhe.*

The saliva has no chemical action on the fats or the albuminous bodies. Its frothiness enables it to carry oxygen into the stomach, and this is thought to be of service. The action of the ptyalin commences with great promptness, and sugar has been detected, it is said, within half a minute after the starch was placed in the mouth. The process, however, is not finished there, but continues after reaching the stomach.— *Valentin.*

The saliva thus prepares a small portion of food for absorption at once, and so insures at the very beginning of the operation of digestion a supply of force-producing material for the immediate use of the system.

‡ We can observe the peculiar motion of the œsophagus by watching a horse's neck when he is drinking.

Fig. 47

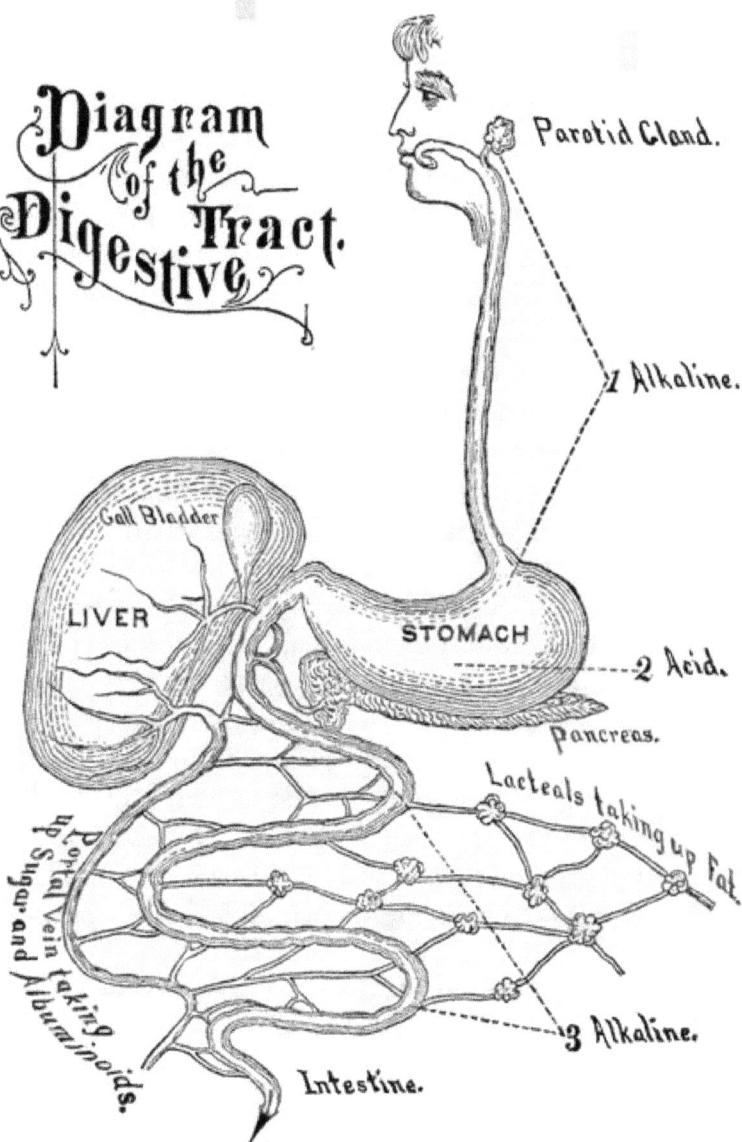

Diagram of the Digestive Tract.

Parotid Gland.

1 Alkaline.

Gall Bladder

LIVER

STOMACH

2 Acid.

Pancreas.

Lacteals taking up Fat

Portal Vein taking up Sugar and Albuminoids.

3 Alkaline.

Intestine.

Diagram of the Digestion of the Food. Notice how the food is submitted to the action of alkaline, acid, and then alkaline fluids. (See note p. 152.)

been compared to that of a bagpipe. It holds about
three pints, though it is susceptible of some distension. It is composed of an inner, mucous membrane, which secretes the digestive fluids ; an outer,
smooth, well-lubricated serous one, which prevents
friction, and between them a stout, muscular coat.
The last consists of two principal layers of longi-
tudinal and circular fibers. When these contract,
they produce a peculiar churning motion, called the
peristaltic (*peri*, round ; *stallein*, to arrange) movement, which thoroughly mixes the contents of the
stomach. At the further end, the muscular fibers
contracting form a gateway, the *pylorus* (a gate), as
it is called, which carefully guards the exit, and
allows no food to pass from the stomach until properly prepared.*

2. THE GASTRIC JUICE.—The lining of the stomach
is soft, velvety, and of a pinkish hue ; but, as soon as
food is admitted, the blood-vessels fill, the surface
becomes of a bright red, and soon there exudes from
the gastric glands a thin, colorless fluid—the gastric
juice. This is secreted to the amount of twelve
pounds per day.† Its acidity is probably due to
muriatic or lactic acid—the acid of sour milk. It
contains a peculiar organic principle called *pepsin* ‡

---

* With a wise discretion, however, it opens for buttons, coins, etc., swallowed
by accident ; when we overload the stomach, it seems to become weary of constantly
denying egress, and, finally, giving up in despair, lets everything through.

† The amount secreted by a healthy adult is variously estimated from five to
thirty-seven pounds. As it is re-absorbed by the blood, there is no loss.

‡ Pepsin is prepared and sold as an article of commerce. The best is said to be
made from the stomachs of young, healthy pigs, which, just before being killed, are
excited with savory food, that they are not allowed to eat. One grain is sufficient
to dissolve 800 grains of coagulated white of egg. A temperature of 130° renders
pepsin inert.

(*peptein*, to digest), which acts as a ferment to produce changes in the food, without being itself modified.

The flow of gastric juice is influenced by various circumstances. Cold water checks it for a time, and ice for a longer period. Anger, fatigue, and anxiety delay and even suspend the secretion. The gastric juice has no effect on the fats or the sugars of the food ; its influence being mainly confined to the albuminous bodies, which it so changes that they become soluble in water.*

The food, reduced by the action of the gastric juice to a grayish. soupy mass, called *chyme* (kīme), escapes through that jealously-guarded door, the pylorus.

Fig. 48.

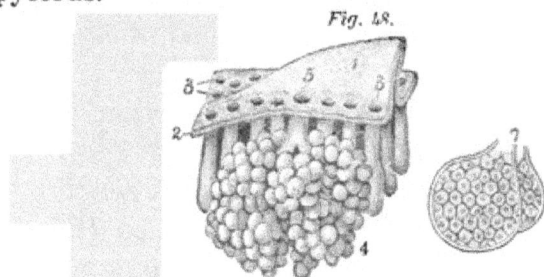

A vertical Section of the Duodenum, highly magnified. 1, *a fold-like villus ;* 2, *epithelium, or cuticle ;* 3. *orifices of intestinal glands ;* 5. *orifice of duodenal glands ;* 4, 7, *more highly magnified sections of the cells of a duodenal gland.*

III. Intestinal Digestion. — The structure of the intestines is like that of the stomach. There is the

---

* The question is often asked why the stomach itself is not digested by the gastric juice, since it belongs to the albuminous substances. Some assign as the reason that life protects that organ, and that living tissues cannot be digested. The fallacy of this has been clearly shown by introducing the legs of live frogs and the ears of rabbits through an opening made in a dog's stomach, where they were readily digested. The latest opinion is that the blood which circulates so freely through the vessels of the lining of the stomach, being alkaline, protects the tissue against the acidity of the gastric juice.

same outer, smooth, serous membrane (peritoneum) to prevent friction, the lining of mucous membrane to secrete the digestive fluids, and the muscular coating to push the food forward. The intestines are divided into the *small*, and the *large*. The first part of the former opens out of the stomach, and is called the *du-o-de'-num*, as its length is equal to the breadth of twelve fingers. Here the chyme is acted upon by the *bile*, and the *pancreatic juice*.

1. THE BILE is secreted by the liver. This gland weighs about four pounds, and is the largest in the body. It is located on the right side, below the diaphragm. The bile is of a dark, golden color, and bitter taste. About three pounds are secreted per day. When not needed for digestion, it is stored in the gall cyst.* Its action on the food, though not fully understood, is necessary to life.†

2. THE PANCREATIC JUICE is a secretion of the pancreas, or "sweet-bread"—a gland nearly as large as the hand, lying behind the stomach. It is alkaline, and contains a ferment called *trypsin*. This juice has the power of changing starch to sugar. Its main work, however, is in breaking up the globules of fat into myriads of minute particles, that mix freely with water, and remain suspended in it like butter in new milk. The whole mass now

---

* A gall-bladder can be obtained from a butcher, and the contents kept in a bottle for examination.

† Experiments have been made with animals by opening the gall-bladder and allowing the gall as secreted to flow out of the body without entering the stomach. Flint describes a case where a dog was thus treated. Although plentifully fed, he died in thirty eight days. He suffered no pain, and death came gradually, merely through a failure of the vital powers.—The alkaline bile, by neutralizing the acid contents of the stomach as they flow into the duodenum, prepares the way for the pancreatic juice, while the bile has also a slight emulsifying power (note p. 154).

Fig. 49.

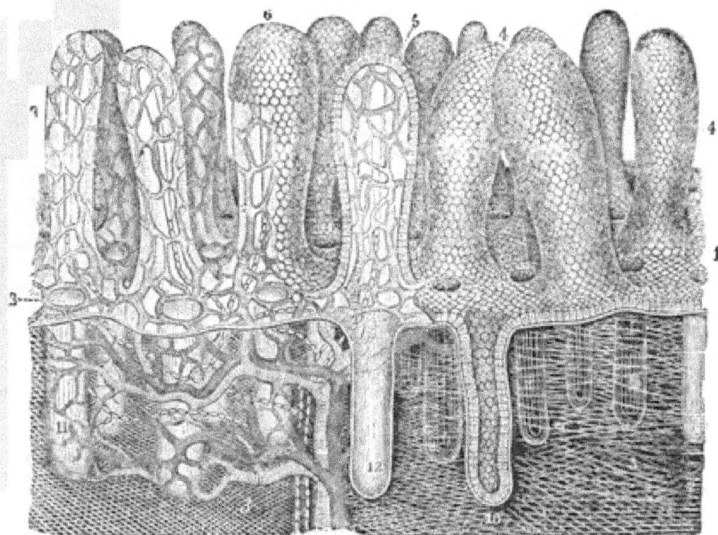

The Mucous Membrane of the Ilium, highly magnified. 1, *cellular structure of the epithelium, or outer layer;* 2, *a vein;* 3, *fibrous layer;* 4, *villi covered with epithelium;* 5, *a villus in section, showing its lining of epithelium, with its blood-vessels and lymphatics;* 6, *a villus partially uncovered;* 7, *a villus stripped of its epithelium;* 8, *lymphatics or lacteals;* 9, *orifices of the glands opening between the villi;* 10, 11, 12, *glands;* 13, *capillaries surrounding the orifices of the gland.*

assumes a milky look, whence it is termed *chyle* (kīle), and passes on to the small intestine.*

3. The Small Intestine is an intricately-folded tube, about twenty feet long, and from an inch to an inch and one-half in diameter. As the chyle passes through this tortuous channel, it receives

---

* "It is curious to observe that while the gastric juice is decidedly acid, the fluids with which the food next comes into contact are alkaline. It is thus submitted to the operation alternately of alkaline, acid, and again of alkaline secretions. In the herbivora there is also a second acid juice. The reason of these alternations is not known, but it can hardly be doubted that they serve to make the digestion of the food more perfect. And although the solvent power of the gastric juice is placed in abeyance when its acidity is neutralized by the alkaline fluids, yet it appears to be the case here, as in respect to the saliva, that effects are produced by the mixture of the various secretions which are poured together into the digestive tube, that would not result from either alone."—*Hinton.*

along the entire route secretions which seem to combine the action of all the previous ones—starch, fat, and albumen being equally affected.

IV. Absorption is performed in two ways, by the *veins*, and the *lacteals*. (1.) The veins in the stomach * immediately begin to take up the water, salt, grape-sugar, and other substances that need no special preparation. The starch, and the albuminous bodies are also absorbed as they are properly digested, and this process continues along the whole length of the alimentary canal. In the small intestine, there is a multitude of tiny projections (*villi*) from the folds of the mucous membrane, more than 7000 to the square inch, giving it a soft, velvety look. These little rootlets, reaching out into the milky fluid, drink into their minute blood-vessels the nutritious part of every sort of food. (2.) The lacteals † (p. 123), a set of vessels starting in the villi side by side with the veins, absorb the principal part of the fat. They convey the chyle through the lymphatics and the thoracic duct (Fig. 43) to the veins, and so within the sweep of the circulation.

The Portal Vein ‡ carries to the liver the food absorbed by the veins of the stomach and the villi of the intestines. On the way, it is greatly modified

---

* The veins and the lacteals are separated from the food by a thin, moist membrane, through the pores of which the fluid-food rapidly passes, in accordance with a beautiful law (*Physics*. page 50) called the *Osmose* of liquids. If two liquids of different densities are separated by an animal membrane, they will mix with considerable force. There is a similar law regulating the interchange of gases through a porous partition, in obedience to which the carbonic acid of the blood, and the oxygen of the lungs, are exchanged through the thin membrane of the air-cells.

† From *lac*, milk, because of the milky look given to their contents by the chyle.

‡ So named because it enters the liver by a sort of gateway.

by the action of the blood itself. In the cells of the
liver, it undergoes as mysterious a process as that
performed by the lymphatic glands, and is then cast
into the circulation.* The food, potent with force,
is now buried in that river of life from which the
body springs momentarily afresh.

The Complexity of the process of digestion, as
compared with the simplicity of respiration and
circulation, is very marked. The mechanical operation of mastication; the lubrication of the food by
mucus; the provision for the security of the respir-
atory organs; the grasping by the muscles of the
throat; the churning movement of the stomach;
the guardianship of the .pylorus; the timely introduction by safe and protected channels of the saliva,
the gastric juice, the bile, the pancreatic juice, and
the intestinal fluids, each with its special adaptation;
the curious peristaltic motion of the intestines; the
twofold absorption by the veins and the lacteals;
the final transformation in the lymphatics, the portal vein, and the liver,—all these present a complexity of detail, the necessity of which can be
explained only when we reflect upon the variety of
the substances we use for food, and the importance
of its thorough preparation before it is allowed to
enter the blood.

The Length of Time Required for digesting a full
meal is from two to four hours. It varies with the

---

* In these cells, the sugar is changed into a kind of starch called *glycogen*. This
is insoluble, and so is stored up in the liver, and even in the substance of the
muscles, until it is needed by the body, when it is once more converted into soluble
sugar and taken up by the circulation. The liver also changes the waste and surplus
albuminous matter into bile, and into urea and uric acid—the forms in which nitro-
genized waste is excreted by the kidneys.

kind of food, state of the system, perfection of mastication, etc. In the celebrated observations made upon Alexis St. Martin* by Dr. Beaumont, his stomach was found empty in two and a half hours after a meal of roast turkey, potatoes, and bread. Pigs' feet and boiled rice were disposed of in an hour. Fresh, sweet apples took one and a half hours; boiled milk, two hours; and unboiled, a quarter of an hour longer. In eggs, which occupied the same time, the case was reversed, — raw ones being digested sooner than cooked. Roast beef and mutton required three, and three and a quarter hours respectively; veal, salt beef, and broiled chicken remained for four hours; and roast pork enjoyed the bad pre-eminence of needing five and a quarter hours.

Value of the Different Kinds of Food.—BEEF and MUTTON possess the greatest nutritive value of any of the meats. LAMB is less strengthening, but more delicate. Like the young of all animals, it should be thoroughly cooked, and at a high temperature, properly to develop its delicious flavor. PORK has much carbon. It sometimes contains a parasite called trichina, which may be transferred to the human system, producing disease and often death.

---

* In 1822, Alexis St. Martin, a Canadian in the employ of the American Fur Company, was accidentally shot in the left side. Two years after, the wound was entirely healed, leaving, however, an opening about two and a half inches in circumference into the stomach. Through this the mucous membrane protruded, forming a kind of valve which prevented the discharge of food but could be readily depressed by the finger, thus exposing the interior. For several years he was under the care of Dr. Beaumont, a skillful physician, who experimented upon him by giving various kinds of food, and watching their digestion through this opening. By means of these observations, and others performed on Katherine Kutt, a woman who had a similar aperture in the stomach, we have very important information as to the digestibility of different kinds of food.

The only preventive is thorough cooking. FISH is
more watery than flesh, and many find it difficult
of digestion. Like meat, it loses its mineral constituents and natural juices when salted, and is
much less nourishing. Oysters are highly nutritious, but are more easily assimilated when raw
than when cooked. MILK is a model food, as it contains albumen, starch, fat, and mineral matter. No
single substance can sustain life for so long a time.
CHEESE is very nourishing—one pound being equal
in value to two of meat, but is not adapted to a
weak stomach. EGGS are most easily digested
when the white is barely coagulated and the yolk is
unchanged. BREAD* should be made of unbolted
flour. The bran of wheat furnishes the mineral
matter we need in our bones and teeth, gives the
bulk so essential to the proper distension of the
organs, and by its roughness gently stimulates them
to action. CORN is rich in fat. It contains, however, more indigestible matter than any other grain,
except oats, and is less nutritious than wheat.†
The POTATO is two-thirds water,—the rest being
mainly starch. RIPE FRUITS, and those vegetables

---

* Very fresh bread, warm biscuit, etc., are condensed by mastication into a pasty
mass that is not easily penetrated by the gastric juice, and hence they are not
healthful. In Germany, bread is not allowed to be sold at the baker's till it is
twenty-four hours old—a wise provision for those who have not strength to resist
temptation. This rule of eating may well be adopted by every one who cares more
for his health than for a gratification of his appetite.

† Persons unaccustomed to the use of corn find it liable to produce derangement
of the digestive organs. This was made fearfully apparent in the prisons at Ander-
sonville during the late civil war. The vegetable food of the Federal prisoners had
hitherto been chiefly wheat-bread and potatoes,—the corn bread so extensively used
at the South being quite new to most of them as a constant article of diet. It soon
became not only loathsome, but productive of serious diseases. On the other hand,
it was the principal article in the rations of the Confederate soldiers, to whom habit
made it a nutritious and wholesome form of food, as was shown by their endurance.
—*Flint's Physiology of Man*, vol. 2, page 41

usually eaten raw, dilute the more concentrated food, and also supply the blood with acids, which are cooling in summer, and useful, perhaps, in assimilation.

The Stimulants.— COFFEE is about half nitrogen, and the rest fatty, saccharine, and mineral substances. It is, therefore, of much nutritive value, especially when taken with milk and sugar. Its peculiar stimulating property is due to a principle called *caffeine*. Its aroma is developed by browning, but destroyed by burning. No other substance so soon relieves the sense of fatigue.* Taken in moderation, it clears the intellect, tranquillizes the nerves, and usually leaves no unpleasant reaction. It serves also as a kind of negative food, since it retards the process of waste.

In some cases, however, it produces a rush of blood to the head, and should be at once discarded. At the close of a full meal it hinders digestion, and at night produces wakefulness. In youth, when the vital powers are strong, and the functions of nature prompt in rallying from fatigue, it is not needed, and may be injurious in stimulating a sensitive organization.

TEA possesses an active principle called *theine*. When used moderately, its effects are similar to those of coffee, except that it rarely produces wakefulness. It contains tannin, which, if the tea is strong, coagulates the albumen of the food—*tans* it— and thus delays digestion. In excess, tea causes

* In the late civil war, the first desire of the soldiers upon halting after a wearisome march, was to make a cup of coffee. This was taken without milk, and often without sugar, yet was always welcome.

nervous tremor, disturbed sleep, palpitation of the heart, and indigestion.*

CHOCOLATE contains much fat, and also nitro-genous matter resembling albumen. Its active principle, *theobromine*,† has some of the properties of caffeine and theine.

The Cooking of Food breaks the little cells, and softens the fibers of which it is composed. In broil-ing or roasting meat, it should be exposed to a strong heat at once, in order to coagulate the albumen upon the outside, and thus prevent the escape of the nutritious juices. The cooking may then be finished at a lower temperature. The same principle applies to boiling meat. In making soups, on the contrary, the heat should be applied slowly, and should reach the boiling point for only a few moments at the close. This prevents the coagulation of the albumen. Frying is an unhealthful mode of cooking food, as thereby the fat becomes partially disorganized.

Rapid Eating produces many evil results. 1. There is not enough saliva mixed with the food ; 2. The coarse pieces resist the action of the digestive fluids ; 3. The food is washed down with drinks that dilute the gastric juice, and hinder its work ; 4. We do not appreciate the quantity we eat until the stomach is overloaded ; 5. Failing to get the taste of our food, we think it insipid, and hence use condi-

* Tea and coffee should be made with boiling water, but should not be boiled afterward. During the "steeping" process, so customary in this country, the vola-tile aroma is lost and a bitter principle extracted. In both England and China it is usual to infuse tea directly in the urn from which it is to be drawn. The tannin in tea is shown when a drop falls on a knife-blade. The black spot is a tannate of iron—a compound of the acid in the tea and the metal.

† It is said that Linnæus the great botanist, was so fond of chocolate that he named the cocoa-tree "Theobroma," the food of the gods.

ments that over-stimulate the digestive organs.   In
these various ways the appetite is depraved, the
stomach vexed, the system overworked, and the
foundation of dyspepsia laid.*

The Quantity and Quality of Food required vary
with the age and habits of each individual.   The
diet of a child† should be largely vegetable, and
more abundant than that of an aged person.   A
sedentary occupation necessitates less food than an
out-door life.   One accustomed to manual labor, on
entering school, should practice self-denial until his
system becomes fitted to the new order of things.
He should not, however, fall into the opposite error.
We read of great men who have lived on bread and
water, and the conscientious student sometimes
thinks that, to be great, he, too, must starve himself.‡   On the contrary, many of the greatest
workers are the greatest eaters.   A powerful engine
needs a corresponding furnace.   Only, we should be
careful not to use more fuel than is needed to run
the machine.

The season should modify our diet.   In winter, we
need highly carbonaceous food, plenty of meat, fat,
etc. ; but in summer we should temper the heat in
our corporeal stoves with fruits and vegetables.

The climate also has its necessities.   The inhabi-

---

* When one is compelled to eat in a hurry, as at a railway station, he would do
well to confine himself principally to meat ; and to dilute this concentrated food
with fruit, crackers, etc., taken afterward more leisurely.

† In youth, repair exceeds waste, and hence the body grows rapidly, and the form
is plump.   In middle life, repair and waste equal each other, and growth ceases.
In old age, waste exceeds repair, and hence the powers are enfeebled and the skin
lies in wrinkles on the shrunken form.

‡ As Dr. Holland well remarks, the dispensation of saw-dust has passed away.
If we desire a horse to win the race, we must give him plenty of oats.

tants of the frigid north have an almost insatiable
longing for fat.*   Thus, in 1812, when the Allies
entered Paris, the Cossacks drank all the oil from
the lamps, and left the streets in darkness.   In trop-
ical regions, a low, unstimulating diet of fruits
forms the chief dependence.†

When Food should be Taken.—On taking food,
the blood sets at once to the alimentary canal, and
the energies are fixed upon the proper performance
of this work.   We should not, therefore, undertake
hard study, labor, or exercise directly after a hearty
meal.   We should give the stomach at least half an
hour.   He who toils with brain or muscle, and thus
centers the blood in any particular organ, before eating, should allow time for the circulation to become
equalized.   There should be an interval of four to
five hours between our regular meals, and there
should be no lunching between times.   With young
children, where the vital processes are more rapid,
less time may intervene.   Nothing should be eaten
within two or three hours of retiring.

* Dr. Hayes, the arctic explorer, says, that the daily ration of the Esquimaux
was from twelve to fifteen pounds of meat, one-third being fat. On one occasion,
he saw a man eat ten pounds of walrus flesh and blubber at a single meal. The low
temperature had a remarkable effect on the members of his own party, and some
of them were in the habit of drinking the contents of the oil-kettle with evident
relish. Other travelers narrate the most incredible stories of the voracity of the
inhabitants of arctic regions. Saritcheff, a Russian admiral, tells of a man who in
his presence ate, at a meal, a mess of twenty-eight pounds of boiled rice and butter,
although he had already partaken of his breakfast. Capt. Cochrane further adds,
in narrating this statement, that he has himself seen three of the savages consume a
reindeer at a sitting.

† A natural appetite for a particular kind of food is an expression not only of
desire but of fitness. Thus the craving of childhood for sugar indicates a need of
the system. It is questionable how far it is proper to force or persuade one to eat
that which he disrelishes, or his stomach loathes. Life within is linked with life
without. Each organ requires its peculiar nutriment, and there is often a peculiar
influence demanded of which we can have no notice except by natural instinct.
Yet, as we are creatures of habit and impulse, we need common sense and good
judgment to correct the too often wayward promptings of an artificial craving.

How Food should be Taken.— A good laugh is the best of sauces. The meal-time should be the happiest hour of the day. Care and grief are the bitter foes of digestion. A cheerful face and a light heart are friends to long life, and nowhere do they serve us better than at the table. God designed that we should enjoy eating, and that, having stopped before satiety was reached, we should have the satisfaction always attendant on a good work well done.

Need of a Variety.— Careful investigations have shown that any one kind of food, however nutritious in itself, fails after a time to preserve the highest working power of the body. Our appetite palls when we confine our diet to a regular routine. Nature demands variety, and she has furnished the means of gratifying it.*

The Wonders of Digestion.—We can understand much of the process of digestion. We can look into the stomach and trace its various steps. Indeed, the chemist can reproduce in his laboratory many of the operations; "a step further," as Fontenelle has said, "and he would surprise nature in the very act." Just here, when he seems so successful,

---

* "She opens her hand, and pours forth to man the treasures of every land and every sea, because she would give to him a wide and vigorous life, participant of all variety. For him the cornfields wave their golden grain—wheat, rye, oats, maize, or rice, each different, but alike sufficing. Freely for him the palm, the date, the banana, the bread-fruit tree, the pine, spread out a harvest on the air; and pleasant apple, plum, or peach solicit his ready hand. Beneath his foot lie stored the starch of the potato, the gluten of the turnip, the sugar of the beet; while all the intermediate space is rich with juicy herbs.

"Nature bids him eat and be merry; adding to his feast the solid flesh of bird, and beast, and fish, prepared as victims for the sacrifice: firm muscle to make strong the arm of toil, in the industrious temperate zone; and massive ribs of fat to kindle inward fires for the sad dwellers under Arctic skies."—*Health and its Conditions.—Hinton.*

he is compelled to pause. At the threshold of life the wisest physiologist reverently admires, wonders, and worships.

How strange is this transformation of food to flesh ! We make a meal of meat, vegetables, and drink. Ground by the teeth, mixed by the stomach, dissolved by the digestive fluids, it is swept through the body. Each organ, as it passes, snatches its particular food. Within the cells of the tissues * it is transformed into the soft, sensitive brain, or the hard, callous bone ; into briny tears. or bland saliva, or acrid perspiration ; bile for digestion, oil for the hair, nails for the fingers. and flesh for the cheek.

Within us is an Almighty Architect, who super-intends a thousand builders, which make in a way past all human comprehension, here a fiber of a muscle, there a filament of a nerve ; here construct-ing a bone, there uniting a tendon,—fashioning each with scrupulous care and unerring nicety. † So, without sound of builder or stroke of hammer, goes up, day by day, the body—the glorious temple of the soul.   .

Diseases, etc. — 1. DYSPEPSIA, or indigestion of food, is generally caused by an over-taxing of the digestive organs. Too much food is used, and the

---

* As the body is composed of individual organs, and each organ of separate tissues, so each tissue is made up of minute cells. Each cell is a little world by itself, too small to be seen by the naked eye, but open to the microscope. It has its own form and constitution as much as a special organ in the body. It absorbs from the blood such food as suits its purposes. Moreover the number of cells in an organ is as constant as the number of organs. As the organs expand with the growth of the body, so the cells of each tissue enlarge, but shrink again with age and the decline of life. Life begins and ends in a cell.—See *Appleton's Cyclopedia*, Art. " Absorption."

† See *Cooke's Religion and Chemistry.* page 236.

entire system is burdened by the excess. Meals are taken at irregular hours, when the fluids are not ready. A hearty supper is eaten when the body, wearied with the day's labor, demands rest. The appetite craves no food when the digestion is enfeebled, but stimulants and condiments excite it, and the unwilling organs are oppressed by that which they cannot properly manage.

Strong tea, alcoholic drinks, and tobacco derange the alimentary function.

Too great variety of dishes, rich food, tempting flavors,—all lead to an overloading of the stomach. This patient, long-suffering member at last wears out. Pain, discomfort, diseases of the digestive organs, and insufficient nutrition are the penalties of violated laws.

2. THE MUMPS are a swelling of the parotid—one of the salivary glands (Fig. 46). The disease is generally epidemic, and the patient should be carefully secluded for the sake of others as well as himself. The swelling may be allowed to take its course. Relief from pain is often experienced by applying flannels wrung out of hot water. Great care should be used not to check the inflammation, and, on first going out after recovery, not to take cold.

ALCOHOLIC DRINKS AND NARCOTICS.

1. ALCOHOL (Continued from p. 134).

Relation of Alcohol to the Digestive Organs.— *Is Alcohol a Food?* To answer this question, let us make a comparison. If you receive into your stomach a piece of bread or beef, Nature welcomes its presence. The juices of the system at once take hold of it, dissolve it, and transform it for the uses of the body. A million tiny fingers (lacteals and villi) reach out to grasp it, work it over, and carry it into the circulation. The blood bears it onward wherever it is needed to mend or to build "The house you live in." Soon, it is no longer bread or beef ; it is flesh on your arm ; its chemical energy is imparted to you, and it becomes your strength.

If, on the other hand, you take into your stomach a little alcohol, it receives no such welcome. Nature treats it as a poison, and seeks to rid herself of the intruder as soon as possible.* The juices of the system will flow from every pore to dilute and weaken it, and to prevent its shriveling up the delicate membranes with which it comes in contact. The veins will take it up and bear it rapidly through the system. Every organ of elimination, all the scavengers of the body—the lungs, the kidneys, the perspiration-glands, at once set to work to

---

* Food is digested. alcohol is not. Food warms the blood. directly or indirectly ; alcohol lowers the temperature. Food nourishes the body, in the sense of assimilating itself to the tissues ; alcohol does not. Food makes blood ; alcohol never does anything more innocent than mixing with it. Food feeds the blood - cells ; alcohol destroys them. Food excites, in health, to normal action only ; alcohol tends to inflammation and disease. Food gives force to the body ; alcohol excites reaction and wastes force. in the first place, and in the second, as a true narcotic, represses vital action and corresponding nutrition.—If alcohol does not act like food, neither does it behave like water. Water is the subtle but innocent vehicle of circulation, which dissolves the solid food, holds in play the chemical and vital reactions of the tissues, conveys the nutritive solutions from cell to cell, from tube to tube, and carries off and expels the effete matter. Water neither irritates tissue, wastes force, nor suppresses vital action : whereas alcohol does all three. Alcohol hardens solid tissue. thickens the blood. narcotizes the nerves, and in every conceivable direction antagonizes the operation and function of water.—*Lees's Text-book.*

throw off the enemy. So surely is this the case, that the breath of a person who has drunk only a single glass of the lightest beer will betray the fact. The alcohol thus eliminated is entirely unchanged.* Nature apparently makes no effort to appropriate it. It courses everywhere through the circulation, and into the great organs, with all its properties unmodified. "Dr. A. B. Hall of Boston states that he once bled a man who was dead drunk. The blood was caught in a bowl, and, on applying a lighted match, the liquid blazed up at once. Experiment shows that to do this it must have contained 20 per cent. of alcohol."

Alcohol, then, is not, like bread or beef, taken hold of, broken up by the mysterious process of digestion, and used by the body. "It cannot therefore be regarded as an aliment," or food.—(*Flint.*) "Beer, wine, and spirits," says Leibig, "contain no element capable of entering into the composition of the blood or the muscular fiber." "That alcohol is incapable of forming any part of the body," remarks Cameron,

---

* Because of the difficulties of such an experiment, we have not yet been able to account satisfactorily by the excretions for all the alcohol taken into the stomach. This remains as yet one of the unsolved problems of physiological chemistry. To collect the whole of the insensible perspiration, for example, is well nigh impossible. It was supposed at one time that a part of the alcohol is oxidized—i. e., burned, in the system. But such a process would impart heat, and it is now proved that alcohol cools, instead of warms, the blood. Moreover, the closest analysis fails to detect in the circulation any trace of the products of alcoholic combustion, such as aldehyde and acetic acid. "The fact," says Flint, "that alcohol is always eliminated, even when drank in minute quantity, and that its elimination continues for a considerable time, gradually diminishing, renders it probable that all that is taken into the body is removed."

The small amount of nutritive substance, chiefly sugar derived from the grain or fruit used in the manufacture of beer or wine, can not, of course, be compared with that contained in bread or beef at the same cost. Leibig says, in his Letters on Chemistry, "We can prove, with mathematical certainty, that as much flour as can lie on the point of a table-knife is more nutritious than eight quarts of the best Bavarian beer."

"is admitted by all physiologists. It cannot be converted into brain, nerve, muscle, or blood."

Effect upon the Digestion.—Experiments tend to prove that alcohol coagulates and precipitates the pepsin from the gastric juice, and so puts a stop to its great work in the process of digestion.*

The greed of alcohol for water causes it to imbibe moisture from the tissues and juices, and to inflame the delicate mucous membrane. It shows the power of nature to adapt herself to circumstances, that the soft, velvety lining of the throat and stomach should come at length to endure the presence of a fiery liquid which, undiluted, would soon shrivel and destroy it. In self-defence, the juices pour in to weaken the alcohol, and it is soon hurried into the circulation. Before this can be done, "it must absorb about three times its bulk of water ;" hence, very strong liquor may be retained in the stomach long enough to interfere seriously with the diges-

---

* The experiments of Dr. Henry Munroe. of Hull. published in the London Medical Journal, are here summarized as showing that the tendency to retard digestion is common to all forms of alcoholic drinks.

FinelyMinced Beef.	2d Hour.	4th Hour.	6th Hour.	8th Hour.	10th Hour.
I. Gastric juice and *water*.	Beef opaque.	Digesting and separating.	Beef much lessened.	Broken up into shreds.	Dissolved like soup.
II. Gastric juice with *alcohol*.	No alteration perceptible.	Slightly opaque, but beef unchanged.	Slight coating on beef.	No visible change.	Solid on cooling. *Pepsin* precipitated.
III. Gastric juice and *pale ale*.	No change.	Cloudy, with fur on beef.	Beef partly loosened.	No further change.	No digestion. *Pepsin* precipitated.

tion, and to injure the lining coat. Habitual use of alcohol permanently dilates the blood-vessels; thickens and hardens the membranes; in some cases, ulcerates the surface; and, finally, "so weakens the assimilation that the proper supply of food cannot be appropriated."—(*Flint.*)*

Effect upon the Liver.—Alcohol is carried by the portal vein directly to the liver. This organ, after the brain, holds the largest share. The influence of the poison is here easily traced. "The color of the bile is soon changed from yellow to green, and even black;" the connective tissue between the lobules becomes inflamed; and, in the case of a confirmed drunkard, hardened and shrunk, the surface often assuming a nodulated appearance known as the "hob-nailed liver." Morbid matter is sometimes deposited, causing what is called "Fatty degeneration," so that the liver is increased to twice or thrice its natural size.

Effect upon the Kidneys.—The kidneys, like the liver, are liable in time to undergo, through the influence of alcohol, a "Fatty degeneration," in which the cells become filled with particles of fat;† the vessels lose their contractility; and, worst of all, the membranes may be so modified as to allow the

* The case of St. Martin (p. 155) gave an excellent opportunity to watch the action of alcohol upon the stomach. Dr. Beaumont summarized his experiments thus: "The free, ordinary use of any intoxicating liquor, when continued for some days, invariably produced inflammation, ulcerous patches, and, finally, a discharge of morbid matter tinged with blood." Yet St. Martin never complained of pain in his stomach, the narcotic influence of the alcohol preventing the signal of danger that Nature ordinarily gives.

† "Disabled by the fatty deposits, the kidneys are unable to separate the waste matter coming to them for elimination from the system. The poisonous material is poured back into the circulation, and often delirium ensues."—*Hubbard.*

albuminous part of the blood to filter through them, and so rob the body of one of its most valuable constituents.*

Does Alcohol Impart Heat?—During the first flush after drinking wine, for example, a sense of warmth is felt. This is due to the tides of warm blood that are being sent to the surface of the body, owing to the "Vascular enlargement" and the rapid pumping of the heart. There is, however, no fresh heat developed. On the contrary, the bringing the blood to the surface causes it to cool faster, reaction sets in, a chilliness is experienced as one becomes sober, and a delicate thermometer placed under the tongue of the inebriate may show a fall of even two degrees below the standard temperature of the body. Several hours are required to restore the usual heat.

As early as 1850, Dr. N. S. Davis, of Chicago, ex-President of the American Medical Association, instituted an extensive series of experiments to determine the effect of the different articles of food and drinks on the temperature of the system. He conclusively proved that, during the digestion of all kinds of food, the temperature of the body is increased, but when alcohol is taken, either in the form of fermented or distilled beverages, the temperature begins to fall within a half-hour, and continues to decrease for two or three hours, and that the reduction of temperature, in extent as well as in duration, is in exact proportion to the amount of alcohol taken.

It naturally follows that, contrary to the accepted opinion, liquor does not fortify against cold. The experience of travelers at the North coincides with that of Dr. Hayes, the Arctic explorer, who says:

* This deterioration of structure frequently gives rise to what is known as "Bright's Disease."—*Richardson*.

" While fat is absolutely essential to the inhabitants and travelers in arctic countries, alcohol is, in almost any shape, not only completely useless, but positively injurious. I have known strong, able-bodied men to become utterly incapable of resisting cold in consequence of the long-continued use of alcoholic drink."

Does Alcohol Impart Strength?—Experience shows that alcohol weakens the power of undergoing severe bodily exertion.* Men who are in training for running, rowing, and other contests where great strength is required, deny themselves all liquors, even when they are ordinarily accustomed to their use.

Dr. Richardson made some interesting experiments to show the influence of alcohol upon muscular contraction. He carefully weighted the hind leg of a frog, and, by means of electricity, stimulating the muscle to its utmost power of contraction, he found out how much the frog could lift. Then administering alcohol, he dis- covered that the response of the muscle to the electrical current became feebler and feebler, as the narcotic began to take effect, until, at last, the animal could raise less than half the amount it lifted by the nat- ural contraction when uninfluenced by alcohol.

Effect upon the Waste of the Body.—The ten- dency of alcohol is to cause a formation of an un- stable substance resembling fat,† and so the use of

* Dr. McRae, in speaking of Arctic exploration. at the meeting of the American Association for the Advancement of Science. held at Montreal in 1856, said : " The moment that a man had swallowed a drink of spirits, it was certain that his day's work was nearly at an end. It was absolutely necessary that the rule of total abstinence be rigidly enforced. if we would accomplish our day's task. The use of liquor as a beverage when we had work on hand, in that terrific cold, was out of the question."

† " The molecular deposits equalizing the waste of the system do not go on regu- larly under the influence of alcohol ; the tissues are not kept up to their standard . and, in time, their composition is changed by a deposit of an amorphous matter resembling fat. This is an unstable substance, and the functions of animal life all retrograde."—*Hubbard on The Opium Habit and Alcoholism.*

liquor for even a short time will increase the weight. But a more marked influence is to check the ordinary waste of the system, so that "the amount of carbonic acid exhaled from the lungs may be reduced as much as 30 to 50 per cent.—(*Hinton.*) The life-process is one of incessant change. Its rapidity is essential to vigor and strength. When the functions are in full play, each organ is being constantly torn down, and as constantly rebuilt with the materials furnished from our food. Anything that checks this oxidation of the tissues, or hinders the deposition of new matter, disturbs the vital functions. Both these results are the inevitable effects of alcohol; for, since the blood contains less oxygen and more carbonic acid, and the power of assimilating the food is decreased, it follows that every process of waste and repair must be correspondingly weakened. The person using liquor consequently needs less bread and beef, and so alcohol seems to him a food—a radical error, as we have shown.

Alcohol Creates a Progressive Appetite for itself. —When liquor is taken, even in the most moderate quantity, it soon becomes necessary, and then arises a craving demand for an increased amount to produce the original effect. No food creates this constantly-augmenting want. A cup of milk drank at dinner does not lead one to go on, day by day, drinking more and more milk, until to get milk becomes the one great longing of the whole being. Yet this is the almost universal effect of alcohol. Hunger is satisfied by any nutritious food : the dram-drinker's thirst demands alcohol. The common experience of mankind teaches us the imminent peril that attends the formation of this progressive poison-habit. A single glass taken as a tonic may lead to the drunkard's grave.

Worse than this, the alcoholic craving may be transmitted from father to son, and young persons often find themselves cursed with a terrible disease known as alcoholism—a keen, morbid appetite for liquor that demands gratification at any cost— stamped upon their very being through the reckless indulgence of this habit on the part of some one of their ancestors.*

The Law of Heredity is, in this connection, well worth consideration. "The world is beginning to perceive," says Francis Galton, "that the life of each individual is, in some real sense, a continuation of the lives of his ancestors." "Each of us is the footing up of a double column of figures that goes back to the first pair." "We are omnibuses," remarks Holmes, "in which all our ancestors ride." We inherit from our parents our features, our physical vigor, our mental faculties, and even much of our moral character. Often, when one generation is skipped, the qualities will reappear in the following one. The virtues, as well as the vices, of our forefathers, have added to, or subtracted from, the strength of our brain and muscle. The evil tendencies of our natures, which it is the struggle of our

* The American Medical Association, at their meeting in St. Paul, Minnesota (1883), *restated* in a series of resolutions their conviction, that "Alcohol should be classed with other powerful drugs ; that when prescribed medically, it should be done with conscientious caution and a sense of great responsibility. That used as a beverage it is productive of a large amount of physical and mental disease ; that it *entails diseased and enfeebled constitutions upon offspring,* and is the cause of a large percentage of the crime and pauperism of our large cities and country."

lives to resist, constitute a part of our heir-looms from the past. Our descendants, in turn, will have reason to bless us only if we hand down to them a pure healthy physical, mental, and moral being.

" There is a marked tendency in nature to transmit all diseased conditions. Thus, the children of consumptive parents are apt to be consumptives. But of all agents, alcohol is the most potent in establishing a heredity that exhibits itself in the destruction of mind and body.* Its malign influence was observed by the ancients long before the production of whisky or brandy, or other distilled liquors, and when fermented liquors or wines only were known. Aristotle says, 'Drunken women have children like unto themselves,' and Plutarch remarks, 'One drunkard is the father of another.' The drunkard by inheritance is a more helpless slave than his progenitor, and his children are more helpless still, unless on the mother's side there is an untainted blood. For there is not only a propensity transmitted, but an actual disease of the nervous system."—*Dr. Willard Parker.*†

PRACTICAL QUESTIONS.

1. How do clothing and shelter economize food ?
2. Is it well to take a long walk before breakfast ?

---

* " Nearly all the diseases springing from indulgence in distilled and fermented liquors are liable to become hereditary, and to descend to at least three or four generations, unless starved out by uncompromising abstinence. But the distressing aspect of the heredity of alcohol is the transmitted drink-crave. This is no dream of an enthusiast, but the result of a natural law. Men and women upon whom this dread inheritance has been forced are everywhere around us, bravely struggling to lead a sober life."—*Dr. Norman Kerr.*

† The subject of alcohol is continued in the chapter on the Nervous System.

3. Why is warm food easier to digest than cold?

4. Why is salt beef less nutritious than fresh? *

5. What should be the food of a man recovering from a fever?

6. Is a cup of black coffee a healthful close to a hearty dinner?

7. Should ice-water be used at a meal?

8. Why is strong tea or coffee injurious?

9. Should food or drink be taken hot?

10. Are fruit-cakes, rich pastry, and puddings wholesome?

11. Why are warm biscuit and bread hard of digestion?

12. Should any stimulants be used in youth?

13. Why should bread be made spongy?

14. Which should remain longer in the mouth, bread or meat?

15. Why should cold water be used in making soup, and hot water in boiling meat?

16. Name the injurious effects of over-eating.

17. Why do not buckwheat cakes, with syrup and butter, taste as well in July as in January?

18. Why is a late supper injurious?

19. What makes a man " bilious "?

20. What is the best remedy? *Ans.* Diet to give the organs rest, and active exercise to arouse the secretions and the circulation.

21. What is the practical use of hunger?

22. How can jugglers drink when standing on their heads?

23. Why do we relish butter on bread?

24. What would you do if you had taken arsenic by mistake? See Appendix.

25. Why should ham and sausage be thoroughly cooked?

26. Why do we wish butter on fish, eggs with tapioca, oil on salad, and milk with rice?

27. Explain the relation of food to exercise.

---

* The French Academicians found that flesh soaked in water so as to deprive it of its mineral matter and juices, lost its nutritive value, and that animals fed on it soon died. Indeed, for all purposes of nutrition, Liebig said it was no better than stones, and the utmost torments of hunger were hardly sufficient to induce them to continue the diet. There was plenty of nutritive food, but there was no medium for its solution and absorption, and hence it was useless.

28. How do you explain the difference in the manner of eating between carnivorous and herbivorous animals?

29. Why is a child's face plump and an old man's wrinkled?

30. Show how life depends on repair and waste.

31. What is the difference between the decay of the teeth and the constant decay of the body?

32. Should biscuit and cake containing yellow spots of soda be eaten?

33. Tell how the body is composed of organs, organs are made up of tissues, and tissues consist of cells.

34. Why do we not need to drink three pints of water per day?

35. Why, during a pestilence, are those who use liquors as a beverage the first, and often the only victims?

36. What two secretions seem to have the same general use?

37. How may the digestive organs be strengthened?

38. Is the old rule, "after dinner sit awhile," a good one?

39. What would you do if you had taken laudanum by mistake? Paris Green? Sugar of lead? Oxalic acid? Phosphorus from matches? Ammonia? Corrosive sublimate? See Appendix.

40. What is the simplest way to produce vomiting, so essential in case of accidental poisoning?

VII.

NERVOUS SYSTEM.

" *Mark then the cloven sphere that holds
All thoughts in its mysterious folds,
That feels sensation's faintest thrill,
And flashes forth the sovereign will;
Think on the stormy world that dwells
Lock'd in its dim and clustering cells;
The lightning gleams of power it sheds
Along its hollow, glassy threads!*"

"*As a king sits high above his subjects upon his throne, and from it
speaks behests that all obey, so from the throne of the brain-cells is all the
kingdom of a man directed, controlled, and influenced. For this occupant,
the eyes watch, the ears hear, the tongue tastes, the nostrils smell, the skin
feels. For it, language is exhausted of its treasures, and life of its experience; locomotion is accomplished, and quiet ensured. When it wills,
body and spirit are goaded like over-driven horses. When it allows, rest
and sleep may come for recuperation. In short, the slightest penetration
may not fail to perceive that all other parts obey this part, and are but
ministers to its necessities.*"—ODD HOURS OF A PHYSICIAN.

THE NERVOUS SYSTEM.

1. THE STRUCTURE.

2. ORGANS OF THE NERVOUS SYSTEM.
- 1. The Brain
  - 1. Description.
  - 2. The Cerebrum.
  - 3. The Cerebellum.
- 2. The Spinal Cord
  - 1. Its Composition.
  - 2. Medulla Oblongata.
- 3. The Nerves
  - 1. Description.
  - 2. Motory and Sensory.
  - 3. Transfer of Pain.
  - 4. The Spinal Nerves— 31 Pairs.
  - 5. The Cranial Nerves— 12 Pairs.
  - 6. Sympathetic System.
  - 7. Crossing of Cords.
  - 8. Reflex Action.
  - 9. Uses of Reflex Action.

3. HYGIENE
- 1. Brain Exercise.
- 2. Connection between Brain-growth and Body-growth.
- 3. Sleep.
- 4. Effect of Sleeping-draughts.
- 5. Sunlight.

4. WONDERS OF THE BRAIN.

5. ALCOHOLIC DRINKS, AND NARCOTICS.
- 1. Alcohol (con't.)
  - 1. Effect of Alcohol upon the Nervous System.
    - 1. Stage of Excitement.
    - 2. Stage of Muscular Weakness.
    - 3. Stage of Mental Weakness.
    - 4. Stage of Unconsciousness.
  - 2. Effect upon the Brain.
  - 3. Effect upon the Mental and the Moral Powers.
- 2. Tobacco.
  - 1. Constituents of Tobacco.
  - 2. Physiological Effects.
  - 3. Possible Disturbances produced by smoking.
  - 4. Influence upon the Nervous System.
  - 5. Is Tobacco a Food ?
  - 6. Influence of Tobacco upon Youth.
- 3. Opium
  - 1. Description.
  - 2. Physiological Effects.
- 4. Chloral Hydrate.
- 5. Chloroform.

THE NERVOUS SYSTEM.*

STRUCTURE.—The nervous system includes the *brain*, the *spinal cord*, and the *nerves*. It is composed of two kinds of matter—the *white*, and the *gray*. The former consists of minute, milk-white, glistening fibers, sometimes as small as $\frac{1}{25,000}$ of an inch in diameter; the latter is made up of small, ashen-colored cells, forming a pulp-like substance of the consistency of blanc-mange.† This is often gathered in little masses, termed ganglions (*ganglion*, a knot), because, when a nerve passes through a group of the cells, they give it the appearance of a knot. The nerve-fibers are conductors, while the gray cells are generators, of nervous force.‡ The

---

* The organs of circulation, respiration, and digestion, of which we have already spoken, are often called the vegetative functions, because they belong also to the vegetable kingdom. Plants have a circulation of sap through their cells corresponding to that of the blood through the capillaries. They breathe the air through their leaves, which act the part of lungs, and they take in food which they change into their own structure by a process which answers to that of digestion. The plant, however, is a mere collection of parts incapable of any combined action. On the other hand, the animal has a nervous system which binds all the organs together.

† In addition to the cells, the gray substance contains also nerve-fibers continuous with the white-fibers, but generally much smaller. These form half the bulk of the gray substance of the spinal cord, and a large part of the deeper layer of the gray matter in the brain.—*Leidy's Anatomy*, p. 507.

‡ What this force is we do not know. In some respects it is like electricity, but, in others, differs materially. Its velocity is about thirty-three metres per second. (*Physics*, p. 183.)

Fig. 50.

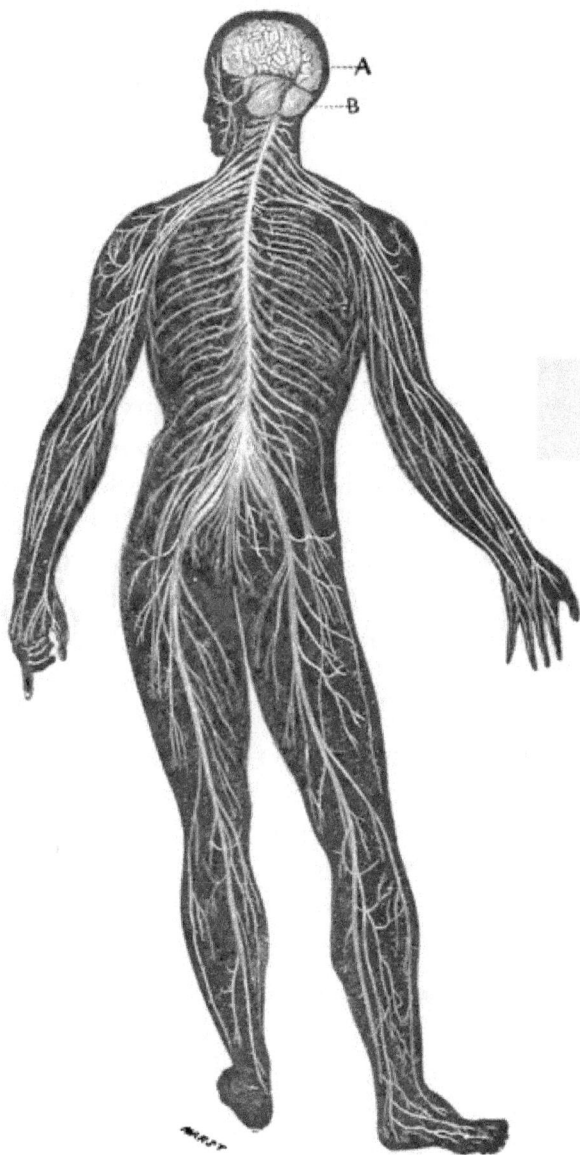

The Nervous System.   A, cerebrum ; B, cerebellum.

ganglia, or nervous centers, answer to the stations
along a telegraphic line, where messages are received
and transmitted, and the fibers correspond to the
wires that communicate between different parts.

The Brain is the seat of the mind.* Its average
weight is about fifty ounces.† It is egg-shaped, and,
soft and yielding, fills closely the cavity of the skull.
It reposes securely on a water-bed, being surrounded
by a double membrane (*arachnoid*), delicate as a
spider's web, which forms a closed sac filled, like the
spaces in the brain itself, with a liquid resembling
water. Within this, and closely investing the brain,
is a fine tissue (*pia mater*), with a mesh of blood-
vessels which dips down into the hollows, and bathes
them so copiously that it uses one-fifth of the entire
circulation of the body. Around the whole is wrapped
a tough membrane (*dura mater*), which lines the bony
box of the skull, and separates the various parts of
the organ by strong partitions. The brain consists
of two parts—the *cerebrum*, and the *cerebellum*.

The Cerebrum fills the front and upper part of the
skull, and comprises about seven-eighths of the entire weight of the brain. As animals rise in the
scale of life, this higher part makes its appearance.

---

* "In proportion to the rest of the nervous matter in the body it is larger in man
than in any of the lower animals. It is the function which the brain performs that
distinguishes man from all other animals, and it is by the action of his brain that he
becomes a conscious, intelligent, and responsible being. The brain is the seat of
that knowledge which we express when we say I. I know it, I feel it, I saw it, are
expressions of our individual consciousness, the seat of which is the brain. It is
when the brain is at rest in sleep that there is least consciousness. The brain may
be put under the influence of poisons, such as alcohol and chloroform, and then the
body is without consciousness. From these and other facts the brain is regarded as
the seat of *consciousness*."—*Lankester*.

† Cuvier's brain weighed 63 ozs.; Webster's, 53½ ozs.; James Fisk's, 58 ozs.;
Ruloff's, 59 ozs.; an idiot's, 19 ozs. See Table in *Flint's Nervous System*.

Fig. 51.

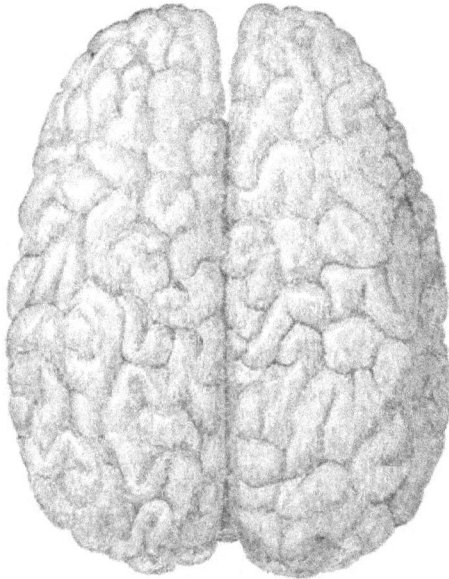

Surface of the Cerebrum.

It is a mass of white fibers, with cells of gray matter
sprinkled on the outside, or lodged here and there in
ganglia. It is so curiously wrinkled and folded as
strikingly to resemble the meat of an English wal-
nut. This structure gives a large surface for the
gray matter,—sometimes as much as 670 square
inches. The convolutions are not noticeable in an
infant, but increase with the growth of the mind,
their depth and intricacy being characteristic of high
mental power.

The cerebrum is divided into two hemispheres,
connected beneath by fibers of white matter. Thus
we have two brains,* as well as two hands and two

* This doubleness has given rise to some curious speculations. In the case of
the hand, eye, etc., we know that the sensation is made more sure. Thus we can see

eyes. This provides us with a surplus of brains, as it were, which can be drawn upon in an emergency. A large part of one hemisphere has been destroyed

Fig. 52.

Pigeon from which the Cerebrum has been removed.

without particularly injuring the mental powers,*— just as a person has been blind in one eye for a long time without having discovered his loss. The cere-

with one eye, but not so well as with both. It is perhaps the same with the brain. We may sometimes carry on a train of thought, " build an air-castle " with one-half of our brain, while the other half looks on and watches the operation ; or, may read and at the same time think of something else. So in delirium, a patient often imagines himself two persons, thus showing a want of harmony between the two halves.—*Draper's Human Physiology*, page 329.

* 'A pointed iron bar, three-and-a-half feet long and one inch and a quarter in diameter, was driven by the premature blasting of a rock completely through the side of the head of a man who was present. It entered below the temple, and made its exit at the top of the forehead, just about the middle line. The man was at first stunned, and lay in a delirious, semi-stupefied state for about three weeks. At the end of sixteen months, however, he was in perfect health, with the wounds healed and with the mental and bodily functions unimpaired, except that the sight was lost in the eye of the injured side."—(*Dalton.*) It is noticeable, however, that the man became changed in disposition, fickle, impatient of restraint, and profane, which he was not before. He died epileptic, probably from progressive disease of the brain, nearly thirteen years after the injury. The tamping-iron and the skull are pre-served in the Warren Anatomical Museum, Boston.

brum is the center of intelligence and thought. Pigeons from which it is removed are plunged in profound stupor, and are inattentive to surrounding objects ; they occasionally open their eyes with a vacant stare, and then relapse into their former apathy.

The Cerebellum lies below the cerebrum, and in the back part of the head (Fig. 50). It is about the size of a small fist. Its structure is similar to that

Fig. 53.

Pigeon from which the Cerebellum has been removed.

of the brain proper, but instead **of con**volutions it has parallel ridges, which, letting the gray matter down deeply into the white matter within, give it a peculiar appearance, called the *arbor-vitæ*, or tree of life (Fig. 55). This part of the brain is the center for the control of the voluntary muscles. Persons in whom it is injured or diseased walk as if intoxicated, and cannot perform any orderly work.

Pigeons from which it is removed are excited, nervous, and try to escape with uncertain, sprawling movements.

The Spinal Cord occupies the cavity of the backbone. It is protected by the same membranes as the brain, but, unlike it, the white matter is on the outside, and the gray matter is within. Deep fissures separate it into halves (Fig. 50), which are, however, joined by a bridge of the same substance. Just as it starts from the brain, there is an expansion called the *medulla oblongata* (Fig. 55).

The Nerves are glistening, silvery threads, composed, like the spinal cord, of white matter without and gray within. They ramify to all parts of the body. Often they are very near each other, yet are perfectly distinct, each conveying its own impression.* Those which carry the orders of the mind to the different organs are called the *motory* nerves; while those which bring back impressions which they receive are styled *sensory* nerves. If the sensory nerve leading to any part be cut, all sensation in that spot will be lost, while motion will remain; if the motory nerve be cut, all motion will be destroyed, while sensation will exist as before.

Transfer of Pain.—Strictly speaking, pain is not in any organ, but in the mind, since only that can feel. When any nerve brings news to the brain of an injury, the mind refers the pain to the end of the nerve. A familiar illustration is seen in the "funny

* Press two fingers together, and, closing the eyes, let some one pass the point of a pin lightly from one to the other; you will be able to tell which is touched, yet if the nerves came in contact with each other anywhere in their long route to the brain, you could not thus distinguish.

bone " behind the elbow. Here the nerve (*ulnar*) gives sensation to the third and fourth fingers, in which, if this bone be struck, the pain will seem to be. Long after a limb has been amputated, pain will be felt in it, as if it still formed a part of the body—any injury in the stump being referred to the point to which the nerve formerly led.*

The nerves are divided into three general classes— the *spinal*, the *cranial*, and the *sympathetic*.

Fig. 54.

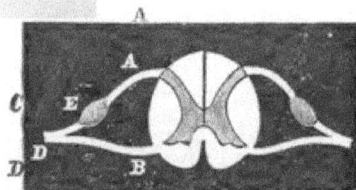

A, *posterior root of a spinal nerve ;* E, *ganglion ;* B, *anterior root ;* D, *spinal nerve.* The white portions of the figure represent the white fibers ; and the dark, the gray.

The Spinal Nerves, of which there are thirty-one pairs, issue from the spinal cord through apertures provided for them in the backbone. Each nerve arises by two roots ; the anterior is the motory, and the posterior the sensory one. The posterior alone connects directly with the gray matter of the cord,

---

* Only about five per cent. of those who suffer amputation lose the feeling of the part taken away. There is something tragical, almost ghastly, in the idea of a spirit limb haunting a man through his life, and betraying him in unguarded moments into some effort, the failure of which suddenly reminds him of his loss. A gallant fellow, who had left an arm at Shiloh, once, when riding, attempted to use his lost hand to grasp the reins while with the other he struck his horse. A terrible fall was the result of his mistake. When the current of a battery is applied to the nerves of an arm-stump, the irritation is carried to the brain, and referred to all the regions of the lost limb. On one occasion a man's shoulder was thus electrized three inches above the point where the limb was cut off. For two years he had ceased to be conscious of his limb. As the current passed through, the man, ignorant of its possible effects, started up, crying, "Oh, the hand ! the hand ! " and tried to seize it with the living grasp of the sound fingers. No resurrection of the dead could have been more startling.—*Dr. Mitchell* on " *Phantom Limbs* " *in Lippincott's Magazine.*

and has a small ganglion of gray matter of its own at a little distance from its origin. These roots soon unite, i. e., are bound up in one sheath, though they preserve their special functions. When the posterior root of a nerve is cut, the animal loses the power of feeling, and when the anterior root is cut, that of motion.

The Cranial Nerves, twelve pairs in number, spring from the lower part of the brain and the medulla oblongata.

Fig. 55.

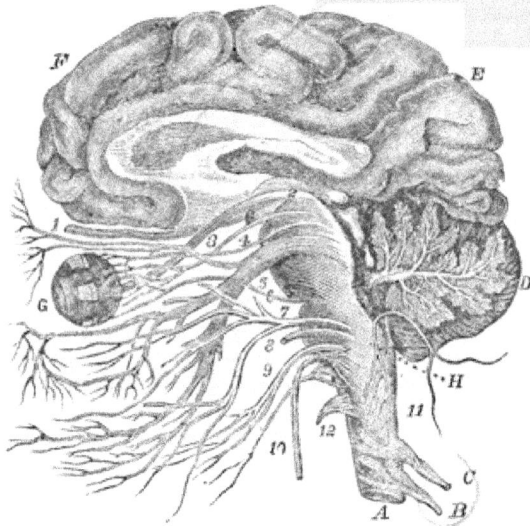

The Brain and the origin of the twelve pairs of Cranial Nerves. F, E, *the cerebrum*; D, *the cerebellum, showing the arbor-vitæ;* G, *the eye;* H, *the medulla oblongata;* A, *the spinal cord;* C *and* B, *the first two pairs of spinal nerves.*

1. The *olfactory,* or first pair of nerves, ramify through the nostrils, and are the nerves of smell.

2. The *optic,* or second pair of nerves, pass to the eyeballs, and are the nerves of vision.

3, 4, 6. The *motores oculi* (eye-movers) are three pairs of nerves used to move the eyes.

5. The *tri-facial*, or fifth pair of nerves, divide each into three branches—hence the name : the first to the upper part of the face, eyes, and nose; the second to the upper jaw and teeth ; the third to the lower jaw and the mouth, where it forms the nerve of taste. These nerves are implicated when we have the toothache or neuralgia.

7. The *facial*, or seventh pair of nerves, are distributed over the face, and give to it expression.*

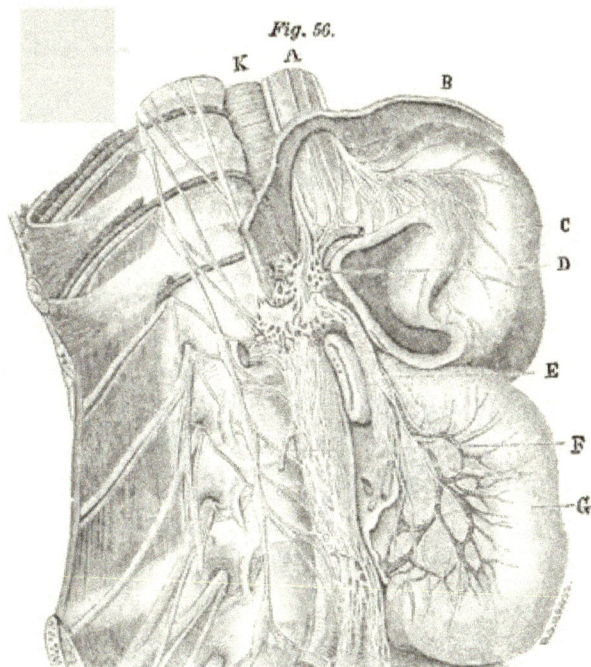

Fig. 56.

Spinal Nerves, Sympathetic Cord, and the Net-work of Sympathetic Nerves around the Internal Organs. K, *aorta;* A, *œsophagus;* B, *diaphragm;* C, *stomach.*

* "If it is palsied, on one side there will be a blank, while the other side will laugh or cry, and the whole face will look funny indeed. There were some cruel people in the middle ages who used to cut the nerve and deform children's faces in this way, for the purpose of making money of them at shows. When this nerve was wrongly supposed to be the seat of neuralgia, or tic-douloureux, it was often cut by surgeons. The patient suffered many dangers, and no relief of pain was gained."— *Mapother.*

8. The *auditory*, or eighth pair of nerves, go to the ears, and are the nerves of hearing.

9. The *glos-so-pha-ryn'-ge-al*, or ninth pair of nerves, are distributed over the mucous membrane of the pharynx, tonsils, etc.

10. The *pneu-mo-gas'-tric*, or tenth pair of nerves, preside over the larynx, lungs, liver, stomach, and one branch extends to the heart. This is the only nerve which goes so far from the head.

11. The *accessory*, or eleventh pair of nerves, rise from the spinal cord, run up to the medulla oblongata, and thence leave the skull at the same opening with the ninth and tenth pairs. They regulate the vocal movements of the larynx.

12. The *hy-po-glos'-sal*, or twelfth pair of nerves, give motion to the tongue.

The Sympathetic System contains the nerves of organic life. It consists of a double chain of ganglia on either side of the backbone, extending into the chest and abdomen. From these, delicate nerves, generally soft and of a grayish color, run to the organs on which life depends—the heart, lungs, stomach, etc.—to the blood-vessels, and to the spinal and cranial nerves over the body. Thus the entire system is bound together with cords of sympathy, so that, "if one member suffers, all the members suffer with it."

Here lies the secret of the control exercised by the brain over all the vital operations. Every organ responds to its changing moods, especially those of respiration, circulation, digestion, and secretion,—processes intimately linked with this system, and controlled by it.

Crossing of Cords.—Each half of the body is presided over, not by its own half of the brain, but that of the opposite side. The motory nerves, as they descend from the brain, in the medulla oblongata, cross each other to the opposite side of the spinal cord. So the motor-nerves of the right side of the body are connected with the left side of the brain, and *vice versa*. Thus a derangement in one half of the brain may paralyze the opposite half of the body. The nerves going to the face do not thus cross, and therefore the face may be motionless on one side, and the limbs on the other. Each of the sensory fibers of the spinal nerves crosses over to the opposite side of the spinal cord, and so ascends to the brain : an injury to the spinal cord may, therefore, cause a loss of motion in one leg and of feeling in the other.

Reflex Action.—Since the gray matter generates the nervous force, a ganglion is capable of receiving an impression, and of sending back or *reflecting* it so as to excite the muscles to action. This is done without the consciousness of the mind.* Thus we wink involuntarily at a flash of light or a threatened blow.† We start at a sudden sound. We jump back

---

* Instances of an unconscious working of the mind are abundant. Abercrombie, in his *Intellectual Powers*, gives the following :

"A lawyer had been excessively perplexed about a very complicated question. An opinion was required from him, but the question was one of such difficulty that he felt very uncertain how he should render it. The decision had to be given at a certain time, and he awoke in the morning of that day with a feeling of great distress. He said to his wife, ' I had a dream, and the whole thing was clearly arranged before my mind, and I would give anything to recover the train of thought.' His wife said to him, ' Go and look on your table.' She had seen him get up in the night and go to his table and sit down and write. He did so, and found there the opinion which he had been most earnestly endeavoring to recover, lying in his own handwriting. There was no doubt about it whatever."

In this case the action of the brain was clearly automatic, i. e.. reflex. The lawyer had worried his brain by his anxiety, and thus prevented his mind from doing its best. But it had received an impulse in a certain direction, and when left to itself, worked out the result. (See Appendix for other illustrations)

† "A very eminent chemist a few years ago was making an experiment upon some extremely explosive compound which he had discovered. He had a small quantity of this compound in a bottle, and was holding it up to the light, looking at it intently ; and whether it was a shake of the bottle or the warmth of his hand, I do

from a precipice before the mind has time to reason upon the danger. The spinal cord conducts certain impressions to the brain, but responds to others without troubling that organ.* The medulla oblongata carries on the process of respiration. The great sympathetic system binds together all the organs of the body.

Uses of Reflex Action.—We breathe eighteen times every minute; we stand erect without a consciousness of effort;† we walk, eat, digest, and at the same time carry on a train of thought. Our brain is thus emancipated from the petty detail of life. If we were obliged to attend to every breath, every pulsation of the heart, every wink of the eye, our time would be wasted in keeping alive. Mere standing would require our entire attention.

not know, but it exploded in his hand, and the bottle was shivered into a million of minute fragments, which were driven in every direction. His first impression was, that they had penetrated his eyes, but to his intense relief he found presently that they had only struck the outside of his eyelids. You may conceive how infinitesimally short the interval was between the explosion of the bottles and the particles reaching his eyes; and yet in that interval the impression had been made upon his sight, the mandate of the reflex action, so to speak, had gone forth, the muscles of his eyelids had been called into action, and he had closed his eyelids before the particles had reached them, and in this manner his eyes were saved. You see what a wonderful proof this is of the way in which the automatic action of our nervous apparatus enters into the sustenance of our lives, and the protection of our most important organs from injury."—*Dr. Carpenter.*

* There is a story told of a man, who, having injured his spinal cord, had lost feeling and motion in his lower extremities. Dr. John Hunter experimented upon him. Tickling his feet, he asked him if he felt it; the man, pointing to his limbs which were kicking vigorously about, answered, "No, but you see my legs do." Illustrations of this independent action of the spinal cord are common in animals. A headless wasp will ply its sting energetically. A fowl, after its head is cut off, will flap its wings and jump about as if in pain, although, of course, all sensation has ceased. "A water beetle, having had its head removed, remained motionless as long as it rested on a dry surface, but when cast into water, it executed the usual swimming motions with great energy and rapidity, striking all its comrades to one side by its violence, and persisting in these for more than half an hour."

† In this way we account for the perilous feats performed by the somnambulist. He is not conscious, as his operations are not directed by the cerebrum, but by the other nervous centers.

Besides, an act which at first demands all our thought soon requires less, and at last becomes mechanical,* as we say, i. e., reflex. Thus we play a familiar tune upon an instrument and carry on a conversation at the same time. All the possibilities of an education and the power of forming habits are based upon this principle. No act we perform ends with itself. It leaves behind it in the nervous centers a tendency to do the same thing again. Our physical being thus conspires to fix upon us the habits of a good or an evil life. Our very thoughts are written in our muscles, so that the expression of our face and even our features grow into harmony with the life we live.

Brain Exercise.—The nervous system demands its life and activity. The mind grows by what it feeds on. One who reads mainly light literature, who lolls on the sofa or worries through the platitudes of an idle or fashionable life, decays mentally; his system loses tone, and physical weakness follows mental poverty. On the other hand, an excessive use of the mind withdraws force from the body, whose weakness, reacting on the brain, produces gradual decay and serious diseases.

The brain grows by the growth of the body. The body grows through good food. fresh air, and work

---

* ' As every one knows," says Huxley, " it takes a soldier a long time to learn his drill—for instance, to put himself into the attitude of ' attention ' at the instant the word of command is heard. But after a time, the sound of the word gives rise to the act, whether the soldier be thinking of it or not. There is a story, which is credible enough. though it may not be true, of a practical joker, who, seeing a discharged veteran carrying home his dinner, suddenly called out ' Attention !' whereupon the man instantly brought his hands down, and lost his mutton and potatoes in the gutter. The drill had been thorough, and its effects had become embodied in the man's nervous structure."

and rest in suitable proportion. For the full development and perfect use of a strong mind, a strong body is essential. Hence, in seeking to expand and store the intellect, we should be equally thoughtful of the growth and health of the body.

Sleep* is as essential as food. During the day, the process of tearing down goes on ; during the night, the work of building up should make good the loss. In youth more sleep is needed than in old age, when nature makes few permanent repairs, and is content with temporary expedients. The number of hours required for sleep must be decided by each person. Napoleon took only five hours, but most people need from six to eight hours,—brain-workers even more. In general, one should sleep until he naturally wakes. If one's rest be broken, it should be made up as soon as possible.

Sunlight.—The influence of the sun's rays upon the nervous system is very marked.† It is said also to

---

* Sleep procured by medicine is rarely as beneficial as that secured naturally. The disturbance to the nervous system is often sufficient to counterbalance all the good results. The habit of seeking sleep in this way, without the advice of a physician, is to be most earnestly deprecated. The dose must be constantly increased to produce the effect, and thus great injury may be caused. Often, too, where laudanum or morphine is used, the person unconsciously comes into a terrible and fatal bondage. (See p. 203.) Especially should infants never be dosed with cordials, as is a common family practice. The damage done to helpless childhood by the ignorant and reckless use of soothing syrups is frightful to contemplate. All the ordinary sleeping-draughts have life-destroying properties, as is proved by the fatal effects of an overdose. At the best, they paralyze the nerve centers, disorder the digestion, and poison the blood. Their promiscuous use is therefore full of danger.

† "The necessity of light for young children is not half-appreciated. Many of their diseases, and nearly all the cadaverous looks of those brought up in great cities, are ascribable to the deficiency of light and air. When we see the glass-room of the photographers in every street, in the topmost story, we grudge them their application to what is often a mere personal vanity. Why should not a nursery be constructed in the same manner? If parents knew the value of light to the skin, especially to children of a scrofulous tendency, we should have plenty of these glass-house nurseries, where children might run about in a proper temperature, free from much of that clothing which at present seals up the skin—that great supplementary

have the effect of developing red disks in the blood.
All vigor and activity come from the sun. Vegetables grown in subdued light have a bleached and
faded look. An infant kept in absolute darkness
would grow into a shapeless idiot. That room is the
healthiest to which the sun has the freest access.
Epidemics frequently attack the inhabitants of the
shady side of a street, and totally exempt those on
the sunny side. If, on a slight indisposition, we
should go out into the open air and bright sunlight,
instead of shutting ourselves up in a close, dark
chamber, we might avoid many a serious illness.
The sun-bath is doubtless a most efficient remedy
for many diseases. Our window blinds and curtains
should be thrown back and open, and we should let
the blessed air and sun stream in to invigorate and
cheer. No house buried in shade, and no room with
darkened windows, is fit for human habitation. In
damp and darkness, lies in wait almost every disease to which flesh is heir. The sun is their only
successful foe.

Wonders of the Brain. — After having seen the
beautiful contrivances and the exquisite delicacy of
the lower organs, it is natural to suppose that when
we come to the brain we should find the most elabo-
rate machinery. How surprising, then, it is to have
revealed to us only cells and fibers! The brain is
the least solid and most unsubstantial looking organ
in the body. Eighty per cent. of water, seven of
albumen, some fat, and a few minor substances

lung—against sunlight and oxygen. They would save many a weakly child who
now perishes from lack of these necessaries of infant life."—*Dr. Winter.*

constitute the instrument which rules the world. Strangest of all, the brain, which is the seat of sensation, is itself without sensation. Every nerve, every part of the spinal cord, is keenly alive to the slightest touch, yet "the brain may be cut, burned, or electrified without producing pain."

ALCOHOLIC DRINKS AND NARCOTICS.

ALCOHOL (Continued from p. 173).

Effect upon the Nervous System.—In the progressive influence of alcohol upon the nervous system, there are, according to the researches of Dr. Richardson, four successive stages.

1. The Stage of Excitement.[*]—The first effect of alcohol, as we have already described on page 130, is to paralyze the nerves that lead to the extreme and minute blood-vessels, and so regulate the passage of the blood through the capillary system. The vital force, thus drawn into the nervous centers, drives the machinery of life with tremendous energy. The heart jumps like the main-spring of a watch when the resistance of the wheels is removed. The blood surges through the body with increased force. Every capillary tube in the system is swollen and flushed, like the reddened nose and cheek.

[*] "The pupil should be careful to note here that alcohol does not act upon the heart directly, and cause it to contract with more force. The idea that alcohol gives energy and activity to the muscles is entirely false. It really, as we have seen (p. 169), weakens muscular contraction. The enfeeblement begins in the first stage and continues in the other stages with increased effect. The heart beats quickly merely because the resistance of the minute controlling vessels is taken off, and it works without being under proper regulation. *What is called a stimulation or excitement is, in absolute fact, a relaxation, a partial paralysis* of one of the most important mechanisms in the animal body. Alcohol should be ranked among the narcotics."—*Richardson.*

In all this there is exhilaration, but no nourishment ; there is animation, but no permanent power conferred on brain or muscle. Alcohol
may cheer for the moment. It may set the sluggish blood in motion,
start the flow of thought, and excite a temporary gayety. "It may enable a wearied or feeble organism to do brisk work for a short time. It
may make the brain briefly brilliant. It may excite muscle to quick
action, but it does nothing at its own cost, fills up nothing it has destroyed, and itself leads to destruction." Even the mental activity it
has excited is an unsafe state of mind, for that just poise of the faculties so essential to good judgment is disturbed by the presence of the
intruder. Johnson well remarked, " Wine improves conversation by
taking the edge off the understanding."

2. The Stage of Muscular Weakness.—If the action of the
alcohol be still continued, the spinal cord is next affected by this
powerful narcotic. The control of some of the muscles is lost. Those
of the lower lip usually fail first, then those of the lower limbs, and the
staggering, uncertain steps betray the result. The muscles themselves, also, become feebler as the power of contraction diminishes.
The temperature, which, for a time, was slightly increased, soon begins
to fall as the heat is radiated ; the body is cooled, and the well-known
"alcoholic chill " is felt.

3. The Stage of Mental Weakness. — The cerebrum is now
implicated. The ideal and emotional faculties are quickened, while the
will is weakened. The center of thought being overpowered, the mind
is a chaos. Ideas flock in thick and fast. The tongue is loosened. The
judgment loses its hold on the acts. The reason giving way, the animal
instincts generally assume the mastery of the man. The hidden nature
comes to the surface. All the gloss of education and social restraint
falls off, and the lower nature stands revealed. The coward shows
himself more craven, the braggart more boastful, the bold more daring,
and the cruel more brutal. The inebriate is liable to become the victim
of any outrage that the slightest provocation may suggest.

4. The Stage of Unconsciousness. — At last, prostration
ensues, and the wild, mad revel of the drunkard ends with utter sense-
lessness. In common speech, the man is "dead drunk." Brain and
spinal cord are both benumbed. Fortunately, the two nervous centers
which supply the heart and the diaphragm are the slowest to be
influenced. So, even in this final stage, the breathing and the circula-

tion still go on, though the other organs have stopped. Were it not for this, every person thoroughly intoxicated would die.*

Effect upon the Brain.—Alcohol seems to have a special affinity for the brain. This organ absorbs more than any other, and its delicate structure is correspondingly affected. The "Vascular enlargement" here reaches its height. The tiny vessels become clogged with blood that is unfitted to nourish, because loaded with carbonic acid, and deprived of the usual quantity of the life-giving oxygen.— (*Hinton.*) The brain is, in the language of the physiologist, malfunctioned. The mind but slowly rallies from the stupor of the fourth stage, and a sense of dullness and depression remains to show with what difficulty the fatigued organ recovers its normal condition. So marked is the effect of the narcotic poison that some authorities hold that "a once thoroughly-intoxicated brain never fully becomes what it was before."

In time, the free use of liquor hardens and thickens the membrane enveloping the nervous matter ; the nerve-corpuscles undergo a "Fatty degeneration"; the blood-vessels lose their elasticity ; and the vital fluid, flowing less freely through the obstructed channels, fails to afford the old-time nourishment.

* Cold has a wonderful influence in hastening this stage, so that a person, previously only in the first stage of excitement, on going outdoors on a winter night, may rapidly sink into a lethargy (become *comatose*), fall, and die. He is then commonly said to have perished with cold. The signs of this coma are of great practical importance, since so many persons die in police stations and elsewhere who are really comatose, when they are supposed to be only sound asleep. The pulse is slow, and almost imperceptible. The face is pale, and the skin cold. "If the arm be pinched it is not moved; if the eyeballs are touched, the lids will not sink." The respiration becomes slower and slower, and, if the person dies, it is because liquid collects in the bronchial tubes, and stops the passage of the air. The man then actually drowns in his own secretions.

The consequent deterioration of the nervous substance—the organ of thought—shows itself in the
weakened mind * that we so often notice in a person
accustomed to drink, and at last lays the foundation
of various nervous disorders — epilepsy, paralysis,
and insanity.† The law of heredity asserts itself
here again, and the inebriate's children often inherit
the disease which he has escaped.

Chief among the consequences of this "perverted
and imperfect nutrition of the brain" is that intermediate state between intoxication and insanity,
well known as Delirium Tremens. "It is characterized by a low, restless activity of the cerebrum,
manifesting itself in muttering delirium, with occasional paroxysms of greater violence. The victim
almost always apprehends some direful calamity;
he imagines his bed to be covered with loathsome
reptiles; he sees the walls of his apartment crowded
with foul specters; and he imagines his friends and
attendants to be fiends come to drag him down to a
fiery abyss beneath."—*Carpenter*.

Influence upon the Mental and Moral Powers.—
So intimate is the relation between the body and the
mind that an injury to one harms the other. The
effect of alcoholized blood is to weaken the will.
The one habitually under its influence often shocks
us by his indecision and breaking his promise to
reform. The truth is, he has lost, in a measure, his
power of self-control. At last, he becomes physically

* "The habitual use of fermented liquors, even to an extent far short of what is
necessary to produce intoxication, injures the body, and diminishes the mental
power."—*Sir Henry Thompson*.

† Casper, the great statistician of Berlin, says: "So far as that city is concerned,
one-third of the insane coming from the poorer classes, were made so by spirit-
drinking."

unable to resist the craving demand of his morbid appetite.

Other faculties share in this mental wreck. The intellectual vision becomes less penetrating, the decisions of the mind less reliable, and the grasp of thought less vigorous. The logic grows muddy. A thriftless, reckless feeling is developed. Ere long, self-respect is lost, and then ambition ceases to allure, and the high spirit sinks.

Along with this mental deterioration comes also a failure of the moral sense. The fine fiber of character undergoes a "degeneration" as certain as that of the muscles themselves. Broken promises tell of a lowered standard of veracity, and a dulled sense of honor, quite as much as of an impaired will. Under the subtle influence of the ever-present poison, signs of spiritual weakness multiply fast. Conscience is lulled to rest. Reason is enfeebled. Customary restraints are easily thrown off. The sensibilities are blunted. There is less ability to appreciate nice shades of right and wrong. Great moral principles and motives lose their power to influence. The judgment fools with duty. The future no longer reaches back its hand to guide the present. The better nature has lost its supremacy.

The wretched victim of appetite will now gratify his tyrannical passion for drink at any expense of deceit or crime. He becomes the blind instrument of his insane impulses, and commits acts from which he would once have shrunk with horror.* Sometimes

* Richardson sums up the various diseases caused by alcohol, as follows: "(a). Diseases of the brain and nervous system, indicated by such names as apoplexy, epilepsy, paralysis, vertigo, softening of the brain, delirium tremens, dipsomania or inordinate craving for drink, loss of memory, and that general failure of the mental

he even takes a malignant pleasure in injuring those whom Nature has ordained he should protect.*

II. TOBACCO.

The Constituents of Tobacco Smoke are numerous, but the prominent ones are carbonic acid, carbonic oxide, and ammonia gases ; carbon, or soot ; and nicotine. The proportion of these substances varies with different kinds of tobacco, the pipe used, and the rapidity of the combustion. Carbonic acid tends to produce sleepiness and headache. Carbonic oxide, in addition, causes a tremulous movement of the muscles, and so of the heart. Ammonia bites the tongue of the smoker, excites the salivary glands, and causes dryness of the mouth and throat. Nicotine is a powerful poison. The amount contained in one or two strong cigars, if thrown directly into the

power, called dementia. (*b*). Diseases of the lungs: one form of consumption, congestion, and subsequent bronchitis. (*c*). Diseases of the heart: irregular beat, feebleness of the muscular walls, dilatation, disease of the valves. (*d*). Diseases of the blood: scurvy, excess of water or dropsy, separation of fibrin. (*e*). Diseases of the stomach: feebleness of the stomach, indigestion, flatulency, irritation, and sometimes inflammation. (*f*). Diseases of the bowels: relaxation or purging, irritation. (*g*). Diseases of the liver: congestion, hardening and shrinking, cirrhosis. (*h*). Diseases of the kidneys: change of structure into fatty or waxy-like condition and other results leading to dropsy, or sometimes to fatal sleep. (*i*). Diseases of the muscles: fatty change in the muscles, by which they lose their power for proper active contraction. (*j*). Diseases of the membranes of the body: thickening and loss of elasticity, by which the parts wrapped up in the membrane are impaired for use, and premature decay is induced."

* It has been argued that a man should not be punished for any crime he may commit during intoxication, but rather for knowingly giving up the reins of reason and conscience, and thus subjecting himself to the rule of his evil passions. Voluntarily to stimulate the mind and put it into a condition where it may drive one to ruin, is very like the act of an engineer who should get up steam in his engine, and then, having opened the valves, desert his post, and let the monster go thundering down the track to sure destruction. Certain persons are thrown into the stage of mental weakness by a single glass of liquor. How can they be excused when the fact of their peculiar liability lends additional force to the argument of abstemiousness, and they know that their only safety lies in total abstinence ?—*Carpenter's Physiology.*

blood, would cause death. Nicotine itself is complex, yielding a volatile substance that gives the
odor to the breath and clothing ; and also a bitter
extract which produces the sickening taste of an
old pipe. In smoking, some of the nicotine is
decomposed, forming pyridine, picoline, and other
poisonous alkaloids.*

Physiological Effects.—The poison of tobacco, set
free by the process either of chewing or smoking,
when for the first time it is swept through the
system by the blood, powerfully affects the body.
Nausea is felt, and the stomach seeks to throw off
the offending substance. The brain is inflamed,
and headache follows. The motor-nerves becoming
irritated, giddiness ensues. Thus Nature earnestly
protests against the formation of this habit. But,
after repeated trials, the system adjusts itself to
the new conditions. A "tolerance" of the poison
is finally established, and smoking causes none of
the former symptoms. Such powerful substances
cannot, however, be constantly inhaled without
producing marked changes. The three great elim-
inating organs—the lungs, the skin, and the kidneys—throw off a large part of the products, but
much remains in the system. When the presence
of the poison is constant, and especially when the
smoking or chewing is excessive, the disturbance
that at first is merely functional, must neces-

---

* The analysis of tobacco as given by different authorities varies greatly. The
one stated in the text suffices for the purposes of this chapter. Von Eulenberg
names several other products of the combustion. One hundred pounds of the dry
leaf may yield as high as seven pounds of nicotine. Havana tobacco contains about
two per cent, and Virginia about six per cent. See *Johnston & Church's Chemistry
of Common Life*, and *Miller's Organic Chemistry*.

sarily, in many cases at least, lead to a chronic derangement.

Probably in this, as in the case of other deleterious articles of diet, the strong and healthy will seem to escape entirely, while the weak and those predisposed to disease will be injured in direct proportion to the extent of the indulgence. Those whose employment leads to active, outdoor work, will show no sign of nicotine poisoning, while the man of sedentary habits will sooner or later be the victim of dyspepsia, sleeplessness, nervousness, paralysis, or other organic difficulties. Even where the user of tobacco himself escapes harm, the law of heredity asserts itself, and the innocent offspring only too often inherit an impaired constitution, and a tendency to nervous complaints.

The Various Disturbances produced in different individuals and constitutions by smoking have been summed up by Dr. Richardson as follows: "(a) In the blood, it causes undue fluidity, and change in the red corpuscles; (b) in the stomach, it gives rise to debility, nausea, and vomiting; (c) in the mucous membrane of the mouth, it produces enlargement and soreness of the tonsils—smoker's sore throat —redness, dryness, and occasional peeling of the membrane, and either unnatural firmness and contraction, or sponginess of the gums; and, where the pipe rests on the lips, oftentimes 'epithelial cancer'; (d) in the heart, it causes debility of the organ, and irregular action; (e) in the bronchial surface of the lungs, when that is already irritable, it sustains irritation, and increases the cough; (f) in the organs of sense, it produces dilation of the pupils of the eye, confusion of vision, bright lines, luminous or cobweb specks, and long retention of images on the retina, with analogous symptoms affecting the ear, viz., inability to define sounds clearly, and the occurrence of a sharp, ringing noise like a whistle; (g) in the brain, it impairs the activity of the organ, oppressing it if it be nourished, but soothing it if it be exhausted; (h) it leads to paralysis in the motor and sympathetic nerves, and to over-secretion from the glands which the sympathetic nerves control."

Is Tobacco a Food?—Here, as in the case of alcohol, the reply is a negative one. Tobacco manifests no characteristic of a food. It cannot impart to the blood an atom of nutritive matter for building up the body. It does not add to, but rather subtracts from, the total vital force. It confers no potential power upon muscle or brain. It stimulates by cutting off the nervous supply from the extremities and concentrating it upon the centers. But stimulation is not nourishment; it is only a rapid spending of the capital stock. There is no greater error than to mistake the exciting of an organ for its strengthening.

The Influence upon Youth.—Here, too, science utters no doubtful voice. Experience asserts only one conviction. *Tobacco retards the development of mind and body.** The law of nature is that of steady growth. It cannot admit of a daily, even though it be merely a functional, disturbance that weakens the digestion, that causes the heart to labor excessively, that prevents the perfect oxidation of the blood, that interferes with the assimilation, and that deranges the nervous system.† No one has a right

* Cigarettes are especially injurious from the irritating smoke of the paper covering, taken into the lungs, and also because the poison-fumes of the tobacco are more directly inhaled. In case of the cheap cigarettes often smoked by boys, the ingredients used are harmful, while one revolts at the thought of the filthy materials, refuse cigar-stumps, &c., employed in their manufacture.

† There is one influence of tobacco that every young man should understand. In many cases, like alcohol, it seems to blunt the sensibilities, and make its user careless of the rights and feelings of others. This is often noticed in common life. We meet occasionally " devotees of the weed," who, ignoring the fact that tobacco is disagreeable to many persons, think only of the gratification of their selfish appetite. They smoke or chew in any place or company. They permit the cigar fumes to blow

thus to check and disturb continually the regular processes of his physical and mental progress. Hence, the young man (especially if he be of a nervous, sensitive organization) who uses tobacco deliberately diminishes the possible energy with which he might commence the work of life ;* while he comes under the bondage of a habit that may become stronger than his will, and under the influence of a narcotic that may beguile his faculties and palsy his strength at the very moment when every power should be awake.

Another peril still lies in the wake of this masterful poison-habit. Tobacco causes thirst and depression that only too often and naturally lead to the use of liquor.

III. OPIUM.

Opium is the dried juice of the poppy. In Eastern countries, this flower is cultivated in immense fields

into the faces of passers-by. They sit where the wind carries the smoke of their pipes so that others must inhale it. They expectorate upon the floor of cars, hotels, and even private homes. They take no pains to remove the odor that lingers about their person and clothing. They force all who happen to be near, their companions, their fellow-travelers, to inhale the nauseating odor of tobacco. Everything must be sacrificed to the one primal necessity of such persons—a smoke. Now, a young man just beginning life, with his fortune to make, and his success to achieve, can illy afford to burden himself with a habit that is costly, that will make his presence offensive to many persons, and that may perhaps render him less sensitive to the best influences, and perceptions of manhood.

* In the Polytechnic School at Paris, the pupils were divided into two classes—the smokers, and the non-smokers. The latter not only excelled on the entrance examinations, but during the entire course of study. Dr. Decaisne examined thirty-eight boys who smoked, and found twenty-seven of them diseased from nicotine poisoning. So long ago as 1868, in consequence of these results, the Minister of Public Instruction forbade the use of tobacco by the pupils.

Dr. Gihon, medical director of the Naval Academy at Annapolis, in his report for 1881, says: "The most important matter in the health-history of the students is that relating to tobacco, and its interdiction is absolutely essential to their future health and usefulness. In this view I have been sustained by my colleagues, and all sanitarians in civil and military life whose views I have been able to obtain."

for the sake of this product. When a cut is made in the poppy-head, a tiny tear of milky juice exudes, and hardens. These little drops are gathered and prepared for the market, an acre yielding, it is said, about twenty-five pounds. Throughout the East, opium is generally smoked ; but in Western countries laudanum and paregoric (tinctures of opium, or morphine—a powerful alkaloid contained in opium), are generally used. The drug itself is also eaten.

Physiological Effect.—Opium, in its various forms, acts directly upon the nerves, a small dose quieting pain, and a larger one soothing to sleep. It arouses the brain, and fires the imagination to a wonderful pitch.* The reaction from this unnatural excitant is correspondingly depressing ; and the melancholy, the "overwhelming horror" that ensues, calls for a renewal of the stimulus. The dose must be gradually increased to produce the original exhilaration.† The seductive nature of the drug leads the unfortunate victim on step by step until he finds himself fast bound in the fetters of the most tyrannical habit known to man.

---

* De Quincy took laudanum for the first time to relieve pain, but such was the intensity of the feelings he then experienced that, as he tells us in his " Confessions of an Opium-eater," " The relief from pain seemed a trifle. Here was a panacea for all human woes. Here was the secret of happiness about which philosophers had disputed for so many ages. Happiness might now be bought for a penny, and carried in the waistcoat pocket : portable ecstasies might be had corked up in a pint bottle."

† " The victim of opium is bound to a drug from which he derives no benefits, but which slowly deprives him of health and happiness, finally to end in idiocy or premature death. Whatever the victim's condition or surroundings may be, the opium must be taken at certain times with inexorable regularity. The liquor or tobacco user can, for a time, go without the use of these agents, and no regular hours are necessary. During sickness, and more especially during the eruptive fevers, he does not desire tobacco or liquor. The opium-eater has no such reprieves ; his dose must

To go on is to wreck all one's powers—physical and mental. To throw off the habit, requires a determination that but few possess. Yet even when the custom is broken, the system is long in recovering from the shock. There seems to be a failure of every organ. The digestion is weakened, food is no longer relished, the muscles waste, the skin shrivels, the nervous centers are paralyzed, and a premature old age comes on apace. De Quincy, four months after he had cast away the opium-bonds, wrote, "Think of me as one still agitated, writhing, throbbing, palpitating, shattered."

No person can be too careful in the use of laudanum, paregoric, and morphine. They may be taken on a physician's prescription as a sedative from racking pain,* but if followed up for any length of time, the powerful habit may be formed ere one is aware. Then comes the opium-eater's grave, or the opium-eater's struggle for life !

IV. CHLORAL HYDRATE.

Chloral Hydrate is a drug frequently used to cause sleep. It leaves behind no headache or lassitude, as is often the case with morphine. It is, how-

---

be taken, and, in painful complications affecting the stomach, a large increase is demanded to sustain the system. If. in forming the habit, two doses are taken each day, the victim is obliged to maintain that number. It is the unceasing, everlasting slavery of regularity that humiliates opium-eaters by a sense of their own weakness. —*Hubbard on The Opium Habit and Alcoholism.*

* Many persons learn to inject laudanum beneath the skin by means of a " hypodermic syringe." The operation is painless, and seems innocent enough. It throws the narcotic directly into the circulation, and relief from pain is often almost instantaneous. But the danger of forming the opium habit is not lessened. and the effect of using the drug in this form for a long time is just as injurious as opium-smoking itself.

ever, a treacherous remedy. It is cumulative in its effects, i. e., even a small and harmless dose, persisted in for a long period, may produce a gradual accumulation of evil results that in the end will prove fatal.

The Physiological Effect is very marked. The appetite becomes capricious. The secretions are unnatural. Nausea and flatulency often ensue. Then the nervous system is involved. The heart is affected. Sleep is broken. Finally, the hydrate being decomposed, by the action of the soda in the blood, into formic acid and chloroform, a new change takes place. The acid combines with the soda, making sodium formate, and the blood, under the influence of this salt, becomes unduly fluid, as it does in the case of persons deprived of fresh food. A disease resembling scurvy is induced, and the skin breaks out in blotches.

V. CHLOROFORM.

Chloroform is a powerful anæsthetic. It is sometimes prescribed by a physician, and afterward (as in the case of laudanum, morphine, and chloral) the sufferer, charmed with the release from pain and the peaceful slumber secured, buys the dangerous drug for himself. Its use soon becomes an apparent necessity. The craving for the narcotic at a stated time is almost irresistible. The patient, compelled to give up the use of the chloroform, will demand, entreat, pray for another dose, in a heart-rending manner, never to be forgotten. Paleness and debility, the earliest symptoms, are followed by mental prostration. Familiarity with the dangerous drug begets carelessness. Its victims are frequently found dead in their beds, with the handkerchief from which they inhaled the volatile poison clutched in their lifeless hands.

PRACTICAL QUESTIONS.

1. Why is the pain of incipient hip-disease frequently felt in the knee?

2. Why does a child require more sleep than an aged person?

3. When you put your finger in the palm of a sleeping child, why will he grasp it?

4. How may we strengthen the brain?

5. What is the object of pain?

6. Why will a blow on the stomach sometimes stop the heart?

7. How long will it take for the brain of a man six feet high to receive news of an injury to his foot, and to reply?

8. How can we grow beautiful?

9. Why do intestinal worms ever affect a child's sight?

10. Is there any indication of character in physiognomy?

11. When one's finger is burned, where is the ache?

12. Is a parlor generally a healthful room?

13. Why can an idle scholar read his lesson and at the same time count the marbles in his pocket?

14. In amputating a limb, what part, when divided, will cause the keenest pain?

15. What is the effect of bad air on nervous people?

16. Is there any truth in the proverb that "he who sleeps dines?"

17. What does a high, wide forehead indicate?

18. How does indigestion frequently cause a headache?

19. What is the cause of one's foot being "asleep"?*

20. When an injury to the nose has been remedied by transplanting skin from the forehead, why is a touch to the former felt in the latter?

21. Are closely-curtained windows healthful?

22. Why, in falling from a height, do the limbs instinctively take a position to defend the important organs?

23. What causes the pylorus to open and close at the right time?

24. Why is pleasant exercise most beneficial?

25. Why does grief cause one to lose his appetite?

26. Why should we never study directly after dinner?

27. What produces the peristaltic movement of the stomach?

28. Why is a healthy child so restless and full of mischief?

29. Why is a slight blow on the back of a rabbit's neck fatal?

30. Why can one walk and carry on a conversation at the same time?

31. What are the dangers of over-study?

32. What is the influence of idleness upon the brain?

33. State the close relation which exists between physical and mental health and disease.

34. In what consists the value of the power of habit?

35. How many pairs of nerves supply the eye?

36. Describe the reflex actions in reading aloud.

37. Under what circumstances does paralysis occur?

38. If the eyelids of a profound sleeper were raised, and a candle brought near, would the iris contract?

39. How does one cough in his sleep?

40. Give illustrations of the unconscious action of the brain.

41. Is chewing tobacco more injurious than smoking?

42. Ought a man to retire from business while his faculties are still unimpaired?

43. Which is the more exhaustive to the brain, worry or severe mental application?

---

* "Here the nervous force is prevented from passing by compression. Just how this is done, or what is kept from passing, we cannot tell. If a current of electricity were moving through a rubber tube full of mercury, a slight squeeze would interrupt it. These cases may depend on the same general principle, but we cannot assert it."—(*Huxley*.) The tingling sensation caused by the compression is transferred to the foot, whence the nerve starts.

44. Is it a blessing to be placed beyond the necessity for work ?*

45. Show how anger, hate, and the other degrading passions are destructive to the brain ? †

46. Are not amusements, to repair the waste of the nervous energy, especially needed by persons whose life is one of care and toil ?

47. Is not severe mental labor incompatible with a rapidly-growing body ?

48. How shall we induce the system to perform all its functions regularly ?

* " It is a poor view of woman's duties and capacities that confines her to a little busy idleness, because the chances of fortune have placed her beyond the necessity of earning a living; and they must have but a narrow notion of the exigencies of social life who believe that any woman of tolerable health and strength may not find abundant opportunities for that kind of work which affords no other recompense than the consciousness of doing good, and therefore to be done by those who can dispense with every other compensation. A life of idleness and luxurious ease can be no more honorable to one sex than to the other, and we know very well that in a man it creates no claims upon the respect and confidence of the community."

† " One of the surest means for keeping the body and mind in perfect health consists in learning to hold the passions in subservience to the reasoning faculties. This rule applies to every passion. Man, distinguished from all other animals by the peculiarity that his reason is placed above his passions to be the director of his will, can protect himself from every mere animal degradation resulting from passionate excitement. The education of the man should be directed not to suppress such passions as are ennobling, but to bring all under governance, and specially to subdue those most destructive passions, anger, hate, and fear."

VIII.

SPECIAL SENSES.

_____

" See how yon beam of seeming white
Is braided out of seven-hued light ;
Yet in those lucid globes no ray
By any chance shall break astray.
Hark, how the rolling surge of sound,
Arches and spirals circling round,
Wakes the hush'd spirit through thine ear
With music it is heaven to hear."

HOLMES.

" Let us remember that if we get a glimpse of the details of natural phenomena, and of those movements which constitute life, it is not in considering them as a whole, but in analyzing them as far as our limited means will permit. In the vibrations of the globe of air which surrounds our planet, as in the undulations of the ether which fills the immensity of space, it is always by molecules which are intangible for us, put in motion by nature, always by the infinitely little, that she acts in exciting the organs of sense, and she has modeled these organs in a proportion which enables them to partake in the movement which she impresses upon the universe. She can paint with equal facility on a fraction of a line of space on the retina, the grandest landscape or the nervelets of a rose-leaf ; the celestial vault on which Sirius is but a luminous point, or the sparkling dust of a butterfly's wing : the roar of the tempest, the roll of thunder, the echo of an avalanche, find equal place in the labyrinth whose almost imperceptible cavities seem destined to receive only the most delicate sounds."

BLACKBOARD ANALYSIS.

THE SPECIAL SENSES.

1. THE TOUCH
- 1. Description of the Organ.
- 2. Its Uses.

2. THE TASTE
- 1. Description of the Organ.
- 2. Its Uses.

3. THE SMELL
- 1. Description of the Organ.
- 2. Its Uses.

4. THE HEARING
- 1. Description of the Organ.
  - a. External Ear.
  - b. Middle Ear.
  - c. Internal Ear.
- 2. How we hear.
- 3. Hygiene of the Ear.

5. THE SIGHT
- 1. Description of the Organ.
- 2. Eyelids, and Tears.
- 3. Structure of the Retina.
- 4. How we see.
- 5. The Use of the Crystalline Lens.
- 6. Near, and Far Sight.
- 7. Color-blindness.
- 8. Hygiene of the Eyes.

THE SPECIAL SENSES.

1. TOUCH.

DESCRIPTION.—Touch is sometimes called the "common sense," since its nerves are spread over the whole body. It is most delicate, however, in the point of the tongue and the tips of the fingers. The surface of the cutis is covered with minute, conical projections called *papillæ* (Fig. 24).* Each of these contains its tiny nerve-twigs, that receive the impression and transmit it to the brain, where the perception is produced.

Uses.—Touch is the first of the senses used by a child. By it we obtain our idea of solidity, and throughout life rectify all other sensations. Thus, when we see anything curious, our first desire is to handle it.

The sensation of touch is generally relied upon. yet, if we hold a marble in the manner shown in Fig. 57, it will seem like two marbles ; and if we touch the fingers thus crossed to our tongue, we shall feel two tongues. Again, if we close our eyes and let another person move one of our fingers over

---

* In the palm of the hand, where there are at least 12,000 in a square inch, we can see the fine ridges along which they are arranged.

a plane surface, first lightly, then with greater pressure, and then lightly again, we shall think the surface concave.

Fig. 57.

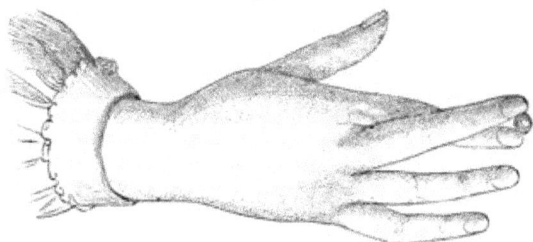

This organ is capable of wonderful cultivation. The physician acquires by practice the *tactus eruditus*,* or learned touch, which is often of great service. The delicacy of touch possessed by the blind almost compensates the loss of the absent sense.†

* An educated sense of touch was possessed by the late Dr. March of Albany, to a remarkable degree. It had been cultivated and perfected in the course of extensive surgical practice to such an extent as to become delicate, precise, and nearly unfailing as a means of diagnosis. On one occasion the Doctor was summoned to see a man who was supposed to have extensive cancerous disease of the thigh, which had been developing many months. The importance of the case had called an unusual number of physicians to the bedside of the patient, and after very thorough examination of the unyielding swelling by them all, Dr. March subjected it to close scrutiny, using his fingers with astonishing caution and delicacy, and appearing to be absorbed in the investigation. A consultation followed in an adjoining room, and from the youngest to the oldest, opinions were expressed that the tumor was malignant in character and called for amputation of the thigh, as affording the only means of arresting the disease, or of saving or prolonging life. Dr. March dissented from this view, and boldly stated that the swelling was an abscess, which could be emptied by a free incision. The reputation of the surgeon, and the positiveness of his assertion, caused his advice to be heeded, and he was permitted to make an attempt to reach the matter, under mental protest of his associates that he would fail in his expectations. The patient, willing to believe almost against hope, suffered the Doctor to proceed with the operation. Taking a scalpel, and guided by his fingers in selecting a location, he made a deep incision through the dense structures of the thigh. Nothing but blood flowed from the wound. A second plunge of the knife brought such an overwhelming discharge of pus from the immensely-distended cavity, as to amount in the quantity collected to several pints. Rapid recovery followed and the limb was saved.—*Dr. Wm. C. Wey.*

† The sympathy between the different organs shows how they all combine to make a home for the mind. When one sense fails, the others endeavor to remedy the defect. It is touching to see how the blind man gets along without eyes, and the

2. TASTE.

Fig. 58.

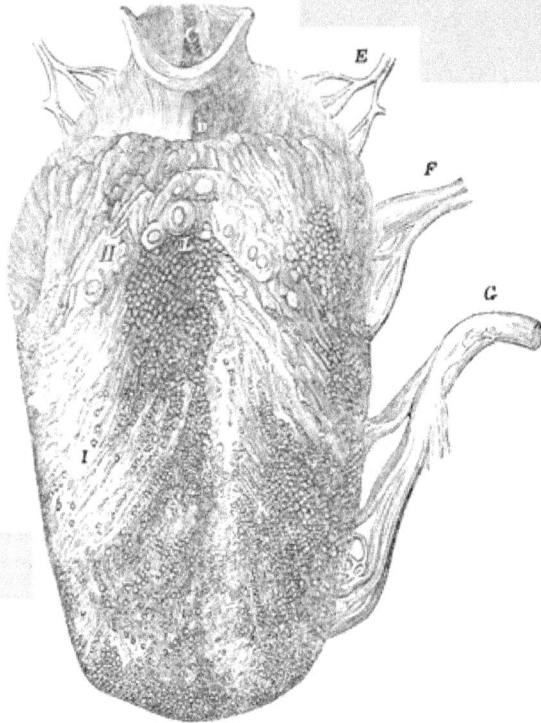

The Tongue, showing the three kinds of Papillæ—the conical (D), the whip-like (K, I), the circumvallate or entrenched (H. L) ; E. F, G, nerves ; C, glottis.—Lankester.

Description.—This sense is located in the papillæ of the tongue and palate. These papillæ start up when tasting, as you can see by placing a drop of vinegar on another person's tongue, or your own, before a

deaf without ears. Cuthbert, though blind, was the most efficient polisher of telescopic mirrors in London. Saunderson, the successor of Newton as professor of mathematics at Cambridge, could distinguish between real and spurious medals. There is an instance recorded of a blind man who could recognize colors. The author knew one who could tell when he was approaching a tree, as he said, by the different feeling of the air.

mirror. The velvety look of this organ is given by
hair-like projections of the cuticle upon some of the
papillæ. They absorb the liquid to be tasted, and
convey it to the nerves.* The back of the tongue is
most sensitive to salt and bitter substances, and, as
this part is supplied by the ninth pair of nerves (Fig.
56), in sympathy with the stomach, such flavors, by
sympathy, often produce vomiting. The edges of
the tongue are most sensitive to sweet and sour substances, and as this part is supplied by the fifth pair
of nerves, which also goes to the face, an acid, by
sympathy, distorts the countenance.

The Use of the Taste was originally to guide in
the selection of food ; but it has become so depraved
by condiments and the force of habit that it would
be a difficult task to tell what are one's natural
tastes.

3. SMELL.†

Description.—The nose, the seat of this sense, is
composed of cartilage covered with muscles and
skin, and joined to the skull by small bones. The
nostrils open at the back into the pharynx, and are
lined by a continuation of the mucous membrane of
the throat. The olfactory nerves (first pair, Fig. 55)
enter through a sieve-like, bony plate at the roof of
the nose, and are distributed over the inner surface
of the two olfactory chambers. The object to be

---

* An insoluble substance is therefore tasteless.

† The sense of smell is so intimately connected with that of taste that we often
fail to distinguish between them.   Garlic, vanilla, coffee, and various spices, which
seem to have such distinct taste, have really a powerful odor but a feeble flavor.

smelled need not touch the nose, but tiny particles
borne on the air enter the nasal passages.*

Fig. 59.

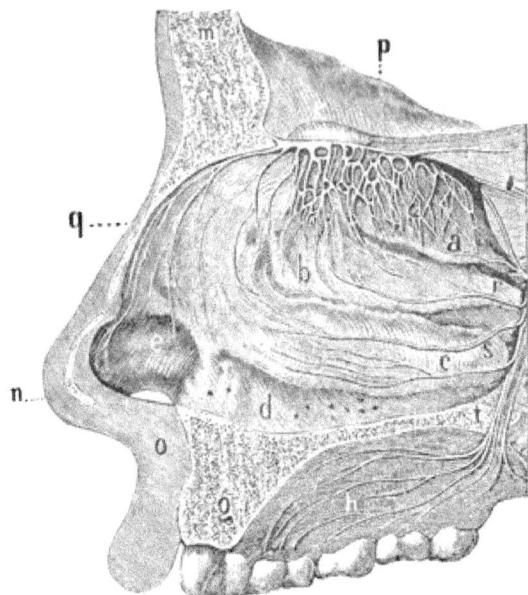

A, b, c, d, *interior of the nose, which is lined by a mucous membrane;* n, *the nose;*
e, *the wing of the nose;* q, *the nose bones;* o, *the upper lip*, g, *section of the upper
jaw-bone;* h, *the upper part of the mouth, or hard palate;* m, *frontal bone of the
skull;* k, *the ganglion or bulb of the olfactory nerve in the skull, from which are seen
the branches of the nerve passing in all directions.*

* "Three-quarters of a grain of musk placed in a room cause a very powerful
smell for a considerable length of time without any sensible diminution in weight,
and the box in which musk has been placed retains the perfume for almost an in-
definite period. Haller relates that some papers which had been perfumed by a grain
of ambergris were still very odoriferous after a lapse of forty years. Odors are
transported by the air to a considerable distance. A dog recognizes his master's
approach by smell even when he is far away; and we are assured by navigators that
the winds bring the delicious odors of the balmy forests of Ceylon to a distance of
ten leagues from the coast. Even after making due allowance for the effects of the
imagination, it is certain that odors act as an excitant on the brain, which may be
dangerous when long continued. They are especially dreaded by the Roman women.
It is well known that in ancient times the women of Rome indulged in a most im-
moderate use of baths and perfumes; but those of our times have nothing in common
with them in this respect; and the words of a lady are quoted, who said on admiring
an artificial rose, ' It is all the more beautiful that it has no smell.' We are warned by
the proverb not to discuss colors or tastes, and we may add odors also.   Men and

The **Uses** of the sense of smell are to guide us in the choice of our food, and to warn us against bad air, and unhealthy localities.

4. HEARING.

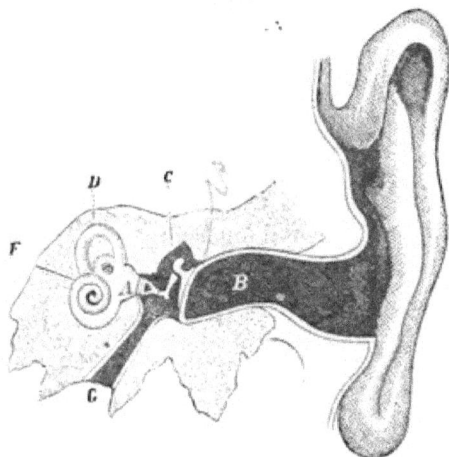

The Ear.

Description.—The ear is divided into the *external, middle,* and *internal* ear.

1. THE EXTERNAL EAR is a sheet of cartilage curiously folded for catching sound. The auditory canal, *B,* or tube of this ear-trumpet, is about an inch long. Across the lower end is stretched *the membrane of the tympanum* or drum, which is kept soft by a fluid wax.

nations differ singularly in this respect. The Laplander and the Esquimaux find the smell of fish-oil delicious. Wrangel says his compatriots, the Russians, are very fond of the odor of pickled cabbage, which forms an important part of their food ; and asafœtida, it is said, is used as a condiment in Persia. and, in spite of its name, there are persons who do not find its odor disagreeable any more than that of vale-rian."—*Wonders of the Human Body.*

2. The Middle Ear is a cavity, at the bottom of which is the Eustachian tube, *G,* leading to the mouth. Across this chamber hangs a chain of three singular little bones, *C,* named from their shape the *hammer,* the *anvil,* and the *stirrup.* All together these tiny bones weigh only a few grains, yet they are covered by a periosteum, are supplied with blood-vessels, and they articulate with perfect joints (one a ball-and-socket, the other a hinge), having synovial membranes, cartilages, ligaments, and muscles.

3. The Internal Ear, or labyrinth, as it is sometimes called from its complex character, is hollowed out of the solid bone. In front, is the vestibule or ante-chamber, *A,* about as large as a grain of wheat; from it open three *semi-circular canals, D,* and the winding stair of the *cochlea,* or snail shell, *E.* Here expand the delicate fibrils of the auditory nerve. Floating in the liquid which fills the labyrinth is a little bag containing hair-like bristles, fine sand, and two ear-stones (*otoliths*). All these knocking against the ends of the nerves, serve to increase any impulse given to the liquid in which they lie. Finally, to complete this delicate apparatus, in the cochlea are minute tendrils, named the fibers of Corti, from their discoverer. These are regularly arranged, — the longest at the bottom, and the shortest at the top. Could this spiral plate, which coils two and a half times around, be unrolled and made to stand upright, it would form a beautiful microscopic harp of three thousand strings. If it were possible to strike these cords as one can the keyboard of a piano, he could produce in the mind

of the person experimented upon every variety of tone which the ear can distinguish.

How We Hear.—Whenever one body strikes another in the air, waves are produced, just as when we throw a stone into the water a series of concentric circles surrounds the spot where it sinks. These waves of air strike upon the membrane. This vibrates, and sends the motion along the chain of bones in the middle ear to the fluids of the labyrinth. Here bristles, sand, and stones pound away, and the wondrous harp of the cochlea, catching up the pulsations,* carries them to the fibers of the auditory nerve, which conveys them to the brain, and gives the mind the idea of sound.

Care of the Ear.—The delicacy of the ear is such that it needs the greatest care. Cold water should not be allowed to enter the auditory canal. If the wax accumulate, never remove it with a hard instrument, lest the delicate membrane be injured, but with a little warm water, after which turn the head to let the water run out, and wipe the ear dry. The hair around the ears should never be left wet, as it may chill this sensitive organ. If an insect get in the external ear, pour in a little oil to kill it, and then remove with tepid water. The object of the Eustachian tube is to admit air into the ear, and thus equalize the pressure on the membrane. If it become closed by a cold, or if, from any cause, the

---

* The original motion is constantly modified by the medium through which it passes. The bristles, otoliths, and Cortian fibers of the ear, and the rods and cones of the eye (p. 221) serve to convert the vibrations into pulsations which act as *stimuli* of the appropriate nerve. The molecular change thus produced in the nerve-fibers is propagated to the brain. (*Physics*, p. 142.)

pressure be made unequal, so as to produce an unpleasant feeling in the ear, relief may often be obtained by grasping the nose and forcibly swallowing.

5 SIGHT.

Fig. 61

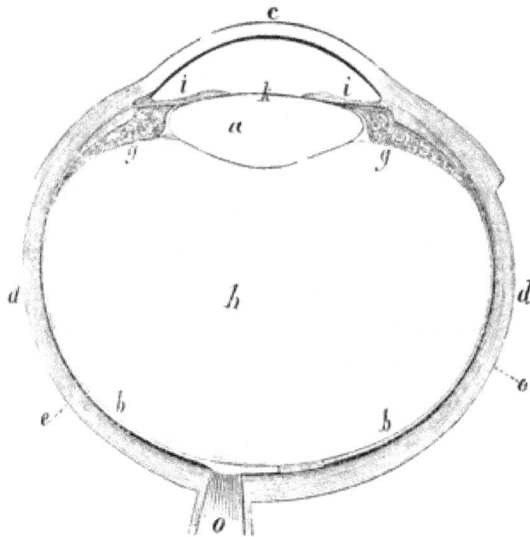

The Eye.

Description.—The eye is lodged in a bony cavity, protected by the overhanging brow. It is a globe, about an inch in diameter. The ball is covered by three coats—(1) the *sclerotic*, *d*, a tough, horny casing, which gives shape to the eye, the convex, transparent part in front forming a window, the *cornea*, *c* ; (2) the *choroid*, *e*, a black lining, to absorb the superfluous light ;* and (3) the *retina*, *b*, a membrane in

---

* Neither white rabbits nor albinos have this black lining, and hence their sight is confused.

which expand fibers of the *optic nerve, o.* The *crystalline lens, a,* brings the rays of light to a focus on the retina. The lens is kept in place by the ciliary processes, *g,* arranged like the rays in the disk of a passion-flower. Between the cornea and the crystalline lens is a limpid fluid termed the *aqueous humor;* while the *vitreous humor*—a transparent, jelly-like liquid—fills the space (*h*) back of the crystalline lens. The pupil, *k,* is a hole in the colored, muscular curtain, *i,* the *iris* (rainbow).

Fig. 62

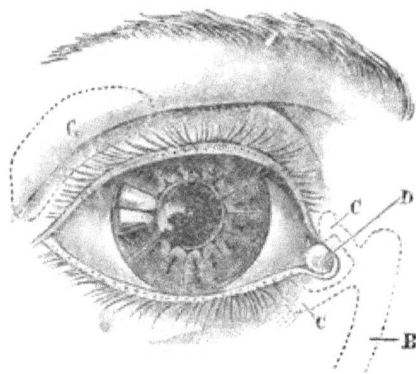

The Eyelashes and the Tear-glands.

Eyelids and Tears.—The eyelids are close-fitting shutters to screen the eye. The inner side is lined with a mucous membrane that is exceedingly sensitive, and thus aids in protecting the eye from any irritating substance. The looseness of the skin favors swelling from inflammation or the effusion of blood, as in a "black eye." The eyelashes serve as a kind of sieve to exclude the dust, and, with the lids, to shield against a blinding light. Just within the lashes are oil glands, which lubricate the edges of the lids, and prevent them from adhering to each other. The tear or *lachrymal* gland, *G*, is an oblong body lodged in the bony wall of the orbit. It empties by several ducts upon the inner surface, at the outer edge of the upper eyelid. Thence the tears, washing the eye, run into the *lachrymal lake*, *D*, a little basin with a rounded border fitted for their reception. On each side of this lake two canals, *C*, *C*, drain off the overplus through the duct. *B*, into the nose. In old age and in disease, these canals fail to conduct the tears away, and hence the lachrymal lake overflows upon the face.

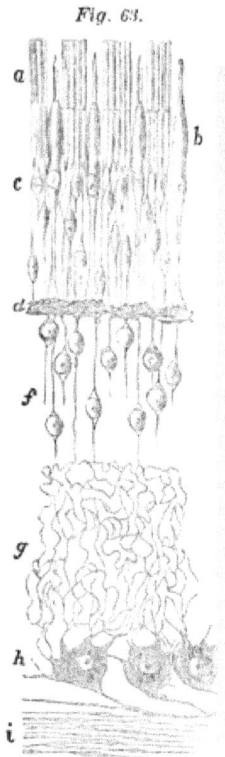

Fig. 63.

Structure of the Retina.

Structure of the Retina.—In Fig. 63 is shown a section of the retina, greatly magnified, since this membrane never exceeds $\frac{1}{80}$ of an inch in thickness. On the inner surface next to the vitreous humor, is a lining membrane not shown in the cut. Next to the choroid and comprising about $\frac{1}{4}$ the entire thickness of the retina, is a multitude of transparent, colorless, microscopic rods, *a*, evenly arranged and packed side by side, like the seeds on the disk of a sunflower. Among them, at regular intervals, are interspersed the cones, *b*. Delicate nerve fibers pass from the ends of the rods and cones, each expanding into a granular body, *c*, thence weaving a mesh, *d*, and again expanding into the granules, *f*. Last is a layer of fine nerve-fibers, *g*, and gray, ganglionic cells, *h*, like the gray matter of the brain, whence filaments extend into *i*, the fibers of the optic nerve.

The layer of rods and cones is to the eye what the bristles, otoliths, and Cortian fibers are to the ear. Indeed, the nerve itself is insensible to light. At the point where it enters the eye, there are no rods

Fig. 64.

and cones, and this is called the *blind spot*. A simple experiment will illustrate the fact. Hold this book directly before the face, and, closing the left eye, look steadily with the right at the left-hand circle in Fig. 64. Move the book back and forth, and a point will be found where the right-hand circle vanishes from sight. At that moment its light falls upon the spot where the rods and cones are lacking.

How We See.—There is believed to be a kind of universal atmosphere, termed *ether*, filling all space. This substance is infinitely more subtle than the air, and occupies its pores, as well as those of all other substances. As sound is caused by waves in the atmosphere, so light is produced by waves in the ether. A lamp-light, for example, sets in motion waves of ether, which pass in through the pupil of the eye to the retina, where the rods and cones transmit the vibration through the optic nerve to the brain, and then the mind perceives the light (note, p. 218).

The Use of the Crystalline Lens. * — A convex lens, as a common burning-glass, bends the rays of

Fig. 65.

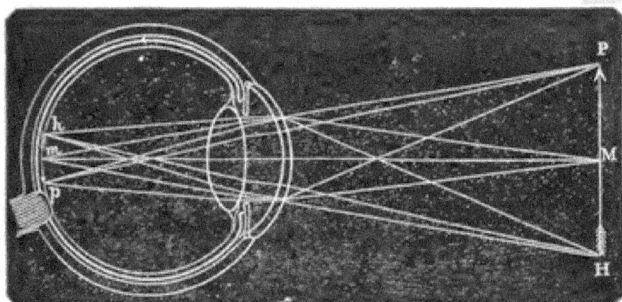

Diagram showing how an image of an object is formed upon the retina by the Crystalline Lens.

light which pass through it, so that they meet at a point called the *focus*. The crystalline lens converges the rays of light which enter the eye, and brings them to a focus on the retina.† The healthy lens has a power of changing its convexity so as to adapt ‡ itself to near and to distant objects. (See Fig. 66.)

---

* The uses of the eye are largely dependent upon the principles of Optics and Acoustics. They are therefore best treated in Physics.

† The cornea and the humors of the eye act in the same manner as the crystalline lens, but not so powerfully.

‡ The simplest way of experimenting on the "adjustment of the eye" is to stick two stout needles upright into a straight piece of wood,—not exactly, but nearly in

Near and Far Sight.—If the lens be too convex, it will bring the rays to a focus before they reach the retina ; if too flat, they will reach the retina before coming to a focus. In either case, the sight will be

Fig. 66.

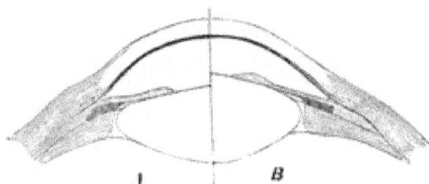

Adjustment of the Crystalline Lens. A, for far objects, and B. for near.

indistinct. A more common defect, however, is in the shape of the globe of the eye, which is either

Fig. 67.

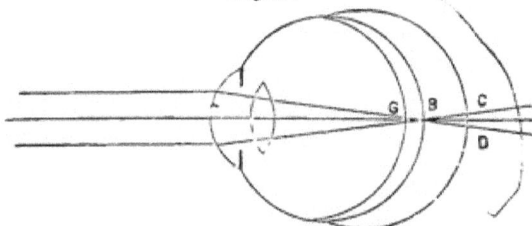

Diagram illustrating the position of the Retina.—B. in natural sight ; G, in far sight ; and C, in near sight.

flattened or elongated. In the former case (see *G*, Fig. 67), objects at a distance can be seen most dis-

the same straight line, so that, on applying the eye to one end of the piece of wood, one needle (*A*) shall be seen about six inches off, and the other (*B*) just on one side of it, at twelve inches distance. If the observer looks at the needle *B* he will find that he sees it very distinctly, and without the least sense of effort: but the image of *A* is blurred and more or less double. Now, let him try to make this blurred image of the needle *A* distinct. He will find he can do so readily enough, but that the act is accompanied by a sense of fatigue. And in proportion as *A* becomes distinct, *B* will become blurred. Nor will any effort enable him to see *A* and *B* distinctly at the same time.—HUXLEY.

tinctly—hence that is called far-sightedness.* In the latter, objects near by are clearer, and hence this is termed near-sightedness. Far-sightedness is remedied by convex glasses; near-sightedness, by concave. When glasses will improve the sight they should be worn ;† any delay will be liable to injure the eyes, by straining their already impaired power. Cataract is a disease in which there is an opacity of the crystalline lens or its capsules, which obscures the vision. The lens may be caused to be absorbed, or may be removed by a skilful surgeon and the defect remedied by wearing convex glasses.

Color-blind Persons receive only two of the three elementary color-sensations (green, red, violet). The spectrum appears to them to consist of two decidedly different colors, with a band of neutral tint between. The extreme red end is invisible, and a bright scarlet and a deep green appear alike. They are unable to distinguish between the leaves of a cherry-tree and its fruit by the color of the two, and see no difference between blue and yellow cloth. Whittier, the poet, it is said, cannot tell red from green unless in direct sunlight. Once he patched some damaged wall-paper in his library by matching a green vine in the pattern with one of a bright autumnal crimson. This defect in the eye is often unnoticed, and many railway accidents have doubt-

---

* This should not be confounded with the long sight of old people, which is caused by the stiffness of the ciliary muscles, whereby the lens cannot adapt itself to the varying distances of objects.

† Dr. Henry W. Williams, the celebrated ophthalmologist, says that, in some ca-es, glasses are more necessary at six or eight years of age than to the majority of healthy eyes at sixty. Sometimes children find accidentally that they can see better through grandmother's spectacles. They should then be supplied with their own.

less happened through an inability to detect the color of signal lights.

Care of the Eyes.—The shape of the eye cannot be changed by rubbing and pressing it, as many suppose, but the sight may thus be fatally injured. Children troubled by near-sightedness should not lean forward at their work, as thereby the vessels of the eye become overcharged with blood. They should avoid fine print, and try, in every possible way, to spare their eyes. If middle age be reached without especial difficulty of sight, the person is comparatively safe. Most cases of squinting are caused by long-sightedness, the muscles being strained in the effort to obtain distinct vision. In childhood, it may be cured by a competent surgeon, who will generally cut the muscle that draws the eye out of place.

Healthy eyes even should never be used to read fine print or by a dim light. Serious injury may be caused by an imprudence of this kind. Reading upon the cars is also a fruitful source of harm. The lens, striving to adapt itself to the incessantly-varying distance of the page, soon becomes wearied.

Objects that get into the eye should be removed before they cause inflammation; rubbing in the mean time only irritates and increases the sensitiveness. If the eye be shut for a few moments, so as to let the tears accumulate, and the upper lid be then lifted by taking hold of it at the center, the cinder or dust is often washed away at once. Trifling objects can be removed by simply drawing the upper lid as far as possible over the lower one; when the lid flies back to its place, the friction will detach any light substance. If it becomes necessary, turn the upper lid over a pencil, and the intruder may then be wiped off with a handkerchief. "Eye-stones" are a popular delusion. When they seem to take out a cinder, it is only because they raise the eyelid, and allow the tears to wash it out. No one should ever use an eye-wash, except by medical advice. The eye is too delicate an organ to be trifled with, and when any disease is suspected, a reliable physician should be consulted. This is especially necessary, since, when one eye is injured, the other, by sympathy, is liable to become inflamed, and perhaps be destroyed.

When reading or working, the *light should be at one side, and never in front.*

The constant increase of defective eyesight among the pupils in our schools is an alarming fact. Dr. Agnew considers that our school-rooms are fast making us a spectacle-using people. Near-sightedness seems to increase from class to class, until, in the upper departments, there are sometimes as high as fifty per cent. of the pupils thus afflicted. The causes are (1), desks so placed as to make the light from the windows shine directly into the eyes of the scholars; (2), cross-lights from opposite windows; (3), insufficient light; (4), small type that strains the eyes; and (5), the position of the pupil as he bends over his desk or slate, causing the blood to settle in his eyes. All these causes can be remedied; the position of the desks can be changed; windows can be shaded, or new ones inserted; books and newspapers that try the eyes can be rejected; and every pupil can be taught how to sit at study.

PRACTICAL QUESTIONS.

1. Why does a laundress test the temperature of her flat-iron by holding it near her cheek ?

2. When we are cold, why do we spread the palms of our hands before the fire?

3. What is meant by a " furred tongue " ?

4. Why has sand or sulphur no taste ?

5. What was the origin of the word palatable?

6. Why does a cold in the head injure the flavor of our coffee ?

7. Name some so-called flavors that are really sensations of touch.

8. What is the object of the hairs in the nostrils?

9. What use does the nose subserve in the process of respiration ?

10. Why do we sometimes hold the nose when we take unpleasant medicine ?

11. Why was the nose placed over the mouth ?

12. Describe how the hand is adapted to be the instrument of touch ?

13. Besides being the organ of taste, what use does the tongue subserve ?

14. Why is not the act of tasting complete until we swallow ?

15. Why do all things have the same flavor when one's tongue is " furred " by fever ?

16. Which sense is the more useful—hearing or sight ?

17. Which coat is the white of the eye ?

18. What makes the difference in the color of eyes ?

19. Why do we snuff the air when we wish to obtain a distinct smell ?

20. Why do red-hot iron and frozen mercury ($-40°$) produce the same sensation ?

21. Why can an elderly person drink tea which to a child would be unbearably hot ?

22. Why does an old man hold his paper so far from his eyes ?

23. Would you rather be punished on the tips of your fingers than on the palm of your hand?

24. What is the object of the eyebrows ?   Are the hairs straight ?

25. What is the use of winking ?

26. When you wink, do the eyelids touch at once along their whole length ?   Why ?

27. How many rows of hairs are there in the eyebrows?

28. Do all nations have eyes of the same shape?

29. Why does snuff-taking cause a flow of tears?

30. Why does a fall cause one to " see stars "?

31. Why can we not see with the nose, or smell with the eyes?

32. What causes the roughness of a cat's tongue?

33. Is the cuticle essential to touch?

34. Can one tickle himself?

35. Why does a bitter taste often produce vomiting?

36. Is there any danger in looking " cross-eyed " for fun?

37. Should school-room desks face a window?

38 Why do we look at a person to whom we are listening attentively?

39. Do we really feel with our fingers?

40. Is the eye a perfect sphere? (See Fig. 61.)

41. How often do we wink?

42. Why is the interior of a telescope or microscope often painted black?

43. What is " the apple of the eye "?

44. What form of glasses do old people require?

45. Should we ever wash our ears with cold water?

46. What is the object of the winding passages in the nose?

47. Can a smoker tell in the dark, whether or not his cigar is lighted?

48. Will a nerve re-unite after it has been cut?

49. Will the sight give us an idea of solidity? *

* " A case occurred a few years ago, in London, where a friend of my own performed an operation upon a young woman who had been born blind, and, though an attempt had been made in early years to cure her, it had failed. She was able just to distinguish large objects, the general shadow, as it were, without any distinct perception of form, and to distinguish light from darkness. She could work well with her needle by the touch, and could use her scissors and bodkin and other implements by the training of her hand, so to speak, alone. Well, my friend happened to see her, and he examined her eyes, and told her that he thought he could get her sight restored ; at any rate, it was worth a trial. The operation succeeded ; and, being a man of intelligence and quite aware of the interest of such a case, he carefully studied and observed it ; and he completely confirmed all that had been previously laid down by the experience of similar cases. There was one little incident which will give you an idea of the education which is required for what you would suppose is a thing perfectly simple and obvious. She could not distinguish by sight the things that she was perfectly familiar with by the touch, at least when they were first presented to her eyes. She could not recognize even a pair of scissors. Now, you

50. Why can a skillful surgeon determine the condition of the brain and other internal organs by examining the interior of the eye ? *

51. Is there any truth in the idea that the image of the murderer can be seen in the eye of the dead victim ?

would have supposed that a pair of scissors, of all things in the world, having been continually used by her, and their form having become perfectly familiar to her hands, would have been most readily recognized by her sight; and yet she did not know what they were; she had not an idea until she was told, and then she laughed, as she said, at her own stupidity. No stupidity at all; she had never learned it, and it was one of those things which she could not know without learning. One of the earliest cases of this kind was related by the celebrated Cheselden, a surgeon of the early part of last century. Cheselden relates how a youth just in this condition had been accustomed to play with a cat and a dog; but for some time after he attained his sight he never could tell which was which, and used to be continually making mistakes. One day, being rather ashamed of himself for having called the cat the dog, he took up the cat in his arms and looked at her very attentively for some time, stroking her all the while ; and in this way he associated the impression derived from the touch, and made himself master (so to speak) of the whole idea of the animal. He then put the cat down, saying, 'Now, puss, I shall know you another time.' "—*Carpenter.*

* This is done by means of an instrument called the ophthalmoscope. Light is thrown into the eye with a concave mirror, and the interior of the organ examined with a lens.

CONCLUSION.

VALUE of Health.—The body is the instrument which the mind uses. If it be dulled or nicked, the effect of the best labor will be impaired. The grandest gifts of mind or fortune are comparatively valueless unless there be a healthy body to use and enjoy them. The beggar, sturdy and brave with his out-door life, is really happier than the rich man in his palace with the gout to twinge him amid his pleasures. The day has gone by when delicacy is considered an element of beauty. Weakness is timid and irresolute; strength is full of force and energy. Weakness walks or creeps ; strength speeds the race, wins the goal, and rejoices in the victory.

False Ideas of Disease.—It was formerly supposed that diseases were caused by evil spirits, who entered the body and deranged its action. Incantations, spells, etc., were resorted to in order to drive them out. By others, disease was thought to come arbitrarily, or as a special visitation of an over-ruling power. Hence, it was to be removed by fasting and prayer. Modern science teaches us that disease is not a thing, but a state. When our food is properly assimilated, the waste matter promptly excreted, and all the organs work in harmony, we are well ;

when any derangement of these functions occurs, we are sick. Sickness is discord, as health is concord. If we abuse or misuse any instrument, we impair its ability to produce a perfect harmony. A suffering body is simply the penalty of violated law.

Prevention of Disease.—Doubtless a large proportion of the ills which now afflict and rob us of so much time and pleasure might easily be avoided. A proper knowledge and observance of hygienic laws would greatly lessen the number of such diseases as consumption, catarrh, gout, rheumatism, dyspepsia, scrofula, etc. There are parts of England where one-half the children die before they are five years old. Every physiologist knows that at least nine-tenths of these lives could be saved by an observance of the simple laws of health. Professor Bennet, in a lecture at Edinburgh, estimated that 100,000 persons die annually in that country from causes easily preventable.

With the advance of science, the causes of many diseases have been determined. Vaccination has been found to prevent or mitigate the ravages of small-pox. Scurvy, formerly so fatal among sailors that it was deemed "a mysterious infliction of Divine Justice against which man strives in vain," is now entirely avoided by the use of vegetables or lime-juice. Cholera, whose approach still strikes dread, and for which there is no known specific, is but the penalty for filthy streets, bad drainage, and over-crowded tenements, and may be controlled, if not prevented, by suitable sanitary measures. It was, no doubt, the intention that we should wear out by the general decay of all the organs,* rather than by the giving out of any single part, and that all should work together harmoniously until the vital force is exhausted.

Cure of Disease.—The first step in the cure of any disease is to obey the law of health which has been violated. If medicine be taken, it is not to destroy the disease, since that is not a thing to be destroyed, but to hold the deranged action in check while nature repairs the injury, and again brings the system into harmonious movement. This tendency of nature is our chief reliance. The best physicians are coming to have diminished confidence in medicine itself, and to place greater dependence upon sanitary and hygienic measures, and the efforts which nature always makes to repair injuries and soothe disordered action. They endeavor only to give her a fair chance, and sometimes to assist her by the intelligent employment of proper medicines. The indiscriminate use of patent nostrums and sovereign remedies of whose constituents we know nothing, and by which powerful drugs are imbibed at haphazard, cannot be too greatly deprecated.† When

---

* "So long as the phenomena of waste and repair are in harmony—so long, in other words, as the builder follows the scavenger—so long man exists in integrity and repair—just, indeed, as houses exist. Derange nutrition, and at once degeneration, or rather let us say, alteration begins. Alas ! that we are so ignorant that there are many things about our house, which, seeing them weaken, we know not how to strengthen. About the brick and the mortar, the frame and the rafters, we are not unlearned; but within are many complexities, many chinks and crannies, full in themselves of secondary chinks and crannies, and these so small, so deep, so recessed, that it happens every day that the destroyer settles himself in some place so obscure, that, while he kills, he laughs at defiance. You or I meet with an accident in our watch. We consult the watchmaker, and he repairs the injury. If we were all that watchmakers, like ourselves, should be, a man could be made to keep time until he died from old age or annihilating accident. This I firmly and fully believe." —*Odd Hours of a Physician.*

† A traveler in Africa states that he was surprised and delighted to find in the

one needs medicine, he needs also a competent physician to advise its use.

Death and Decay.—By a mystery we cannot understand, life is linked with death, and out of the decay of our bodies they, day by day, spring afresh. At last the vital force which has held death and decay in bondage, and compelled them to minister to our growth, and serve the needs of our life, faints and yields the struggle. These powers which have so long time been our servants, gather about our dying couch, and their last offices usher us into the new life and the grander possibilities of the world to come. This last birth, we who see the fading, not the dawning, life, call death.

> " O Father! grant Thy love divine,
> To make these mystic temples Thine,
> When wasting age and wearying strife
> Have sapp'd the leaning walls of life ;
> When darkness gathers over all,
> And the last tottering pillars fall,
> Take the poor dust Thy mercy warms,
> And mold it into heavenly forms."
>
> *Holmes.*

possession of the chief medicine man of one of the interior tribes a carefully preserved copy of the New York *Tribune.* On inquiry, he found that it was exceedingly valuable, as a minute fragment of it either rubbed on the outside or taken inwardly was a sovereign remedy for as long a list of diseases as ever graced the advertisement of an American pain-killer. The mania which some people possess for tippling with patent medicines is no more sensible than the trust of the poor savage in a New York daily.

APPENDIX.

H I N T S

T H E   S I C K - R O O M.

A SICK-ROOM should be the lightest and cheeriest in the house.
A small, close, dark bedroom or a recess is bad enough for one
in health, but unendurable for a sick person. In a case of fever, and
in many acute diseases, it should be remote from the noise of the
family; but when one is recovering from an accident, and in all
attacks where quiet is not needed, the patient may be where he can
amuse himself by watching the movements of the household, or looking out upon the street.

The ventilation must be thorough. Bad air will poison the sick and
the well alike. A fireplace is, therefore, desirable. Windows should
open easily. By carefully protecting the patient with extra blankets,
the room may be frequently aired. If there be no direct draught, much
may be done to change the air, by simply swinging an outer door to
and fro many times.

A bare floor, with strips of carpet here and there to deaden noise, is
cleanest, and keeps the air freest from dust. Cane-bottomed chairs
are preferable to upholstered ones. All unnecessary furniture should
be removed out of the way. A straw bed or a mattress is better than
feathers. The bed-hangings, lace curtains, etc., should be taken down.
Creaking hinges should be oiled. Sperm candles are better than
kerosene lamps.

Never whisper in a sick-room. All necessary conversation should be
carried on in the usual tone of voice. Do not call a physician
unnecessarily, but if one be employed *obey his directions* implicitly.

Never give nostrums over-officious friends may suggest. Do not allow visitors to see the patient, except it be necessary. Never bustle about the room, nor go on tip-toe, but move in a quiet, ordinary way. Do not keep the bottles in the continued sight of the sick person. Never let drinking-water stand in the room.

Do not raise the patient's head to drink, but have a cup with a long spout, or use a bent tube, or even a straw. Do not tempt the appetite when it craves no food. Bathe frequently, but let the physician prescribe the method. Give written directions to the watchers. Have all medicines carefully marked. Remove all soiled clothing, etc., at once from the room. Change the linen much oftener than in health. When you wish to change the sheets, and the patient is unable to rise, roll the under sheet tightly lengthwise to the middle of the bed ; put on the clean sheet, with half its width folded up, closely to the other roll ; lift the patient on to the newly-made part, remove the soiled sheet, and then spread out the clean one.

DISINFECTANTS.

An excellent disinfectant may be made by dissolving in a pail of water any one of the three following : (1) a fluid ounce of carbolic acid ; (2) half an ounce of permanganate of potash ; (3) a pound of green vitriol. The solution of the first kind may be sprinkled on the floor or on the bedding, or allowed to evaporate in the room. Bedding may be washed in the solution of the second substance. Ill odors in the sick-room will be instantly removed by evaporating a few drops of carbolic acid. Vaults, drains, etc., may be purified by the solution of the third kind. Chloride of lime may be used for the same purpose.

WHAT TO DO TILL THE DOCTOR COMES.

Burns.—When a person's clothes catch fire, quickly lay him on the ground, wrap him in a coat, mat, shawl, carpet, or in his own clothes, as best you can to extinguish the fire. Pour on plenty of water till the half-burned clothing is cooled. Then carry the sufferer to a warm room, lay him on a table or a carpeted floor, and with a sharp knife or scissors remove his clothing.

The treatment of a burn consists in protecting from the air.* An excellent remedy is to apply soft cloths kept wet with sweet oil, or *cold water which contains all the " cooking soda" that it will dissolve.* Afterward dress the wound with carbolic acid salve. Wrap a dry bandage upon the outside. Then remove the patient to a bed and warmly cover.† Apply cold water to a small burn till the smart ceases, and then cover with ointment. Do not remove the dressings until they become stiff and irritating ; then take them from a part at a time ; dress and cover again quickly.

Cuts, Wounds, etc.—The method of stopping the bleeding has been described on page 126. If an artery is severed, a physician should be called at once. If the bleeding is not profuse, apply cold water until it ceases, dry the skin, draw the edges of the wound together, and secure them by strips of adhesive plaster. Protect with an outer bandage. This dressing should remain for several days. In the meantime wet it frequently with cold water to subdue inflammation. When suppuration begins, wash occasionally with tepid water and Castile soap.

Dr. Woodbridge, of New York, in a recent address, gave the following directions as to " What to do in case of a sudden wound when the surgeon is not at hand." "An experienced person would naturally close the lips of the wound as quickly as possible, and apply a bandage. If the wound is bleeding freely, but no artery is spouting blood, the first thing to be done is to wash it with water at an ordinary temperature. To every pint of water add either five grains of corrosive sublimate, or two and a half teaspoonfuls of carbolic acid. If the acid is used, add two table-spoonfuls of glycerine, to prevent its irritating the wound. If there is neither of these articles in the house, add four table-spoonfuls of borax to the water. Wash the wound, close it, and apply a compress of a folded square of cotton or linen. Wet it in the

* It is a great mistake to suppose that salves will " draw out the fire " of a burn, or heal a bruise or cut. The vital force must unite the divided tissue by the deposit of material. and the formation of new cells.

† If a burn be near a joint or on the face, even if small. let a doctor see it. and do not be in any hurry about having it healed. Remember that with all the care and skill which can be used, contractions will sometimes take place. The danger to life from a burn or scald is not in proportion to its severity, but to its extent—that is, a small part, such as a hand or a foot or a face, may be burned so deeply as to cripple it for life, and yet not much endanger the general health ; but a slight amount of burning, a mere scorching, over two-thirds of the body, may prove fatal.—*Hope.*

solution used for washing the wound, and bandage down quickly and firmly. If the bleeding is profuse, a sponge dipped in very hot water and wrung out in a dry cloth should be applied as quickly as possible. If this is not available, use ice, or cloths wrung out in ice water. If a large vein or artery is spouting, it must be stopped at once by compression. This may be done by a rubber tube wound around the arm tightly above the elbow or above the knee, where the pulse is felt to beat ; or an improvised 'tourniquet' may be used. A hard apple or a stone is placed in a folded handkerchief, and rolled firmly in place. This bandage is applied so that the hard object rests on the point where the artery beats, and is then tied loosely around the arm. A stick is thrust through the loose bandage and turned till the flow of blood ceases."

Bleeding from the Nose is rarely dangerous, and often beneficial. When it becomes necessary to stop it, sit upright and compress the nostrils between the thumb and forefinger, or with the thumb press upward upon the upper lip. A piece of ice, a snow-ball, or a compress wet with cold water may be applied to the back of the neck.

A Sprain is often more painful and dangerous than a dislocation. Wrap the injured part in flannels wrung out of hot water, and cover with a dry bandage, or, better, with oiled silk. Liniments and stimulating applications are injurious in the first stages, but useful when the inflammation is subdued. *Do not let the limb hang down.* It must be kept quiet, even after all pain has ceased. If used too quickly, dangerous consequences may ensue.

Diarrhea, Cholera Infantum, etc., are often caused by eating indigestible food or by checking of the perspiration ; but more frequently by peculiar conditions of the atmosphere, especially in large cities. If the limbs are cold, give a hot bath, and rub thoroughly. *If possible, go to bed and lie quietly on the back. Rest is better than medicine.* If there be pain, apply repeatedly to the abdomen flannels wrung out of hot water. If medicine is needed, take fifteen drops of peppermint and thirty of paregoric in a wine-glass of warm water ; or an adult may take twenty drops of spirits of camphor and thirty to forty drops of laudanum. Laudanum should rarely be given to an infant, except by a physician's order. Eat no fruit, vegetables, pastry, or pork. If much thirst exist, give small pieces of ice, or cold tea or toast-water.

Croup.—Send at once for a doctor. Induce vomiting by syrup of ipecac or mustard and water. Put the feet in a hot bath. Apply hot fomentations rapidly renewed to the chest and throat.

Sore Throat.—Wrap the neck in a wet bandage, and cover with red flannel or a woolen stocking. Gargle the throat frequently with a solution of a tea-spoonful of salt in a pint of water, or thirty grains of chlorate of potash in a wine-glass of water.

Fits, Apoplexy, Epilepsy, etc.—Loosen the clothing, and raise the head and shoulders, but do not bend the head forward on the neck. Apply cold to the head, and heat to the feet. Follow with an emetic. In a child, a full hot bath is excellent. When there are convulsions, prevent the patient from injuring himself; especially put something in his mouth to keep him from biting his tongue.

Toothache and Earache.—Insert in the hollow tooth, or in the ear, cotton wet with laudanum, spirits of camphor, or chloroform. When the nerve is exposed, wet it with creosote or carbolic acid. Hot cloths or a hot brick wrapped in cloth and held to the face will often relieve the toothache. In a similar manner treat the ear, wetting the cloth in hot water, and letting the vapor pass into the ear.

Choking.—Ordinarily a smart blow between the shoulders, causing a compression of the chest and a sudden expulsion of the air from the lungs, will throw out the substance. If the person can swallow, and the object be small, give plenty of bread or potato, and water to wash it down. Press upon the tongue with a spoon, when, perhaps, you may see the offending body, and draw it out with a blunt pair of scissors. If neither of these remedies avail, give an emetic of syrup of ipecac or mustard and warm water.

Frost Bites are frequently so sudden that one is not aware when they occur. In Canada it is not uncommon for persons meeting in the street to say, "Mind, sir, your nose looks whitish." The blood cools and runs slowly, and the blood-vessels become choked and swollen. *Keep from the heat.* Rub the part quickly with snow, if necessary for hours, till the natural color is restored. If one is benumbed with cold, take him into a cold room, remove the wet clothes, rub the body dry, cover with blankets, and give a little warm tea or weak brandy and water. On recovering, let him be brought to a fire gradually.*

Fevers, and many acute diseases, are often preceded by a loss of appetite, headache, shivering, "pains in the bones," indisposition to

---

* If you are caught in a snow-storm, look for a snow-bank in the lee of a hill, or a wood out of the wind, or a hollow in the plain filled with snow. Scrape out a hole big enough to creep into, and the drifting snow will keep you warm. Men and animals have been preserved after days of such imprisonment. Remember that if you give way to sleep in the open field, you will never awake.

work, etc. In such cases, sponge with tepid water, and rub the body till all aglow. Go to bed, place hot bricks to the feet, take nothing but a little gruel or beef tea, and drink moderately of warm, cream-of-tartar water. If you do not feel better the next morning, call a physician. If that be impossible, take a dose of castor-oil or Epsom salt.

Sun-stroke is a sudden prostration caused by intense heat. The same effect is produced by the burning rays of the sun and the fierce fire of a furnace. When a person falls under such circumstances, place your hand on his chest. If the skin be cool and moist, it is not a sun-stroke; but if it be dry and "biting hot," there can be no mistake. Time is now precious. At once carry the sufferer to the nearest pump or hydrant, and dash cold water on the head and chest until consciousness is restored.—*Dr. H. C. Wood.*

To prevent sun-stroke, wear a porous hat, and in the top of it place a wet handkerchief; also drink freely of water, not ice cold, to induce abundant perspiration.

Asphyxia, or apparent death, whether produced by drowning, suffocation, bad air, or coal gas, requires very similar treatment. Send at once for blankets, dry clothing, and a physician. Treat the sufferer upon the spot, if the weather be not too unfavorable.

1. Loosen the clothing about the neck and chest.

2. Turn the patient on his face, open the mouth, draw out the tongue, and cleanse the nostrils, so as to clear the air-passages.

3. Place the patient on his back, grasp his arms firmly above the elbows, and pull them gently upward until they meet over the head, in order to draw air into the lungs. Then bring the arms back by the side, to expel the air. Repeat the process about fifteen times per minute. Alternate pressure upon the chest, and blowing air into the mouth through a quill or with a pair of bellows, may aid your efforts. Excite the nostrils with snuff or smelling salts, or by passing hartshorn under the nose. Do not cease effort while there is hope. Life has been restored after five hours of suspended animation.

4. When respiration is established, wrap the patient in dry, warm clothes, and rub the limbs under the blankets or over the dry clothing energetically *toward the heart.* Apply heated flannels, bottles of hot water, etc., to the limbs, and mustard plasters* to the chest.

Foreign Bodies in the Ear.—Insects may be killed by dropping a little sweet oil into the ear. Beans, peas, matches, etc., may gener-

---

* The best mustard poultice is the paper plaster now sold by every druggist. It is always ready, and can be carried by a traveler. It has only to be dipped in water, and applied at once.

ally be removed by *cautiously* syringing the ear out with tepid water. Do not use much force lest the tympanum be injured. If this fail, dry the ear, stick the end of a little linen swab into thick glue, let the patient lie on one side, put this into the ear until it touches the substance, keep it there three-quarters of an hour while it hardens, and then draw them all out together. Be careful that the glue does not touch the skin at any point, and that you are at work upon the right ear. Children often deceive one as to the ear which is affected.

Foreign Bodies in the Nose, such as beans, cherry-pits, etc., may be frequently removed by closing the opposite nostril, and then blowing into the child's mouth forcibly. The air, unable to escape except through the other nostril, will sweep the obstruction before it.

ANTIDOTES TO POISONS.

Acids : *Nitric* (aqua fortis), *hydrochloric* (muriatic), *sulphuric* (oil of vitriol), *oxalic*, etc.—Drink a little water to weaken the acid, or, still better, take strong soap-suds. Stir some magnesia in water, and drink freely. If the magnesia be not at hand, use chalk, soda, lime, whiting, soap, or even knock a piece of plaster from the wall, and scraping off the white outside coat pound it fine, mix with milk or water, and drink at once. Follow with warm water, or flax-seed tea.

Alkalies : *Potash, soda, ley, ammonia* (hartshorn).—Drink weak vinegar or lemon juice. Follow with castor or linseed oil, or thick cream.

Antimony : *Antimonial Wine, tartar emetic*, etc.—Drink strong, green tea, and in the mean time chew the dry leaves. The direct antidote is a solution of nut-gall or oak-bark.

Arsenic : *Cobalt, Scheele's green, fly-powder, ratsbane*, etc.—Give *plenty of milk, whites of eggs*, or induce vomiting by mustard and warm water, or even soap-suds.

Bite of a Snake or a Mad Dog. — Tie a bandage above the wound, if on a limb. Wash the bite thoroughly, and, if possible, let the person suck it strongly. Rub some lunar caustic or potash in the wound, or heat the point of a small poker or a steel-sharpener white hot, and press it into the bite for a moment. It will scarcely cause pain, and will be effectual in arresting the absorption of the poison, unless a vein has been struck.

Copper : *Sulphate of copper* (blue vitriol), *acetate of copper* (verdigris). —Take whites of eggs or soda. Use milk freely.

Laudanum : *Opium, paregoric, soothing cordial, soothing syrup*, etc.— Give an emetic at once of syrup of ipecac, or mustard and warm water, etc. After vomiting, use strong coffee freely. *Keep the patient awake* by pinching, pulling the hair, walking about, dashing water in the face, and any expedient possible.

Lead : *White lead, acetate of lead* (sugar of lead), *red lead*. Give an emetic of syrup of ipecac, or mustard and warm water, or salt and water. Follow with a dose of Epsom salt.

Matches : *Phosphorus.*—Give magnesia, chalk, whiting, or even flour in water, and follow with mucilaginous drinks.

Mercury · *Calomel, chloride of mercury* (corrosive sublimate, bug poison), *red precipitate.*—Drink milk copiously. Take the whites of eggs, or even stir flour in water, and use freely.

Nitrate of Silver (lunar caustic).—Give salt and water, and follow with castor-oil.

Nitrate of Potash (salpetre, nitre).—Give mustard and warm water, or syrup of ipecac. Follow with flour and water, and cream or sweet oil.

Prussic Acid (oil of bitter almonds), *cyanide of potassium.*—Take a tea-spoonful of hartshorn in a pint of water. Apply smelling salts to the nose, and dash cold water in the face.

Sting of an Insect.—Apply a little hartshorn or spirits of cam phor, or soda moistened with water, or a paste of clean earth and saliva.

Sulphate of Iron (green vitriol).—Give syrup of ipecac. or mustard and warm water. or any convenient emetic; then magnesia and water.

QUESTIONS FOR CLASS USE.

The questions include the notes. The figures refer to the pages.

INTRODUCTION

ILLUSTRATE the value of physiological knowledge. Why should physiology be studied in youth? When are our habits formed? How do habits help us? Why should children prize the lessons of experience. How does Nature punish a violation of her laws? Name some of Nature's laws. What is the penalty of their violation? Name some bad habits and their punishments. Some good habits and their rewards. How do the young ruin their health? Compare one's constitution with a deposit in the bank. Can one in youth lay up health as he can money for middle or old age? (See *Conclusion.*) Is not the preservation of one's health a moral duty? What is suicide?

THE SKELETON.

How many bones are there in the body? Is the number fixed? Is the length of the different bones proportional? What is an organ? A function? Name the three uses of the bones. Why do the bones have such different shapes? Why are certain bones hollow? Round? Illustrate.

6. What is the composition of bone? How does it vary? How can you remove the mineral matter? The animal matter? Why is a burned bone white and porous? What is the use of each of the constituents of a bone? What food do dogs find in bones? What is " bone black "?

7. What is ossification? Why are not the bones of children as

easily broken as those of aged persons? Why do they unite so much quicker? What are the fontanelles? Describe the structure of a bone.

8. What is the object of the filling? Why does the amount vary in different parts of a bone? What is the appearance of a bone seen through a microscope? What is the periosteum? Is a bone once removed ever restored?

9. What are the lacunæ? The Haversian canals? Why so called? *Ans.* From their discoverer, Havers. Define a bone.* What occupies the lacunæ? *Ans.* The bone-cells (osteoblasts). How do bones grow? Illustrate. How does a broken bone heal? How rapidly is bone produced? Illustrate.

10. Objects of "splints"? Describe how a joint is packed. Lubricated. How are the bones tied together? What is a tissue? Illustrate.

11. Name the three general divisions of the bones. What is the object of the skull? Which bone is movable? How is the lower jaw hinged?

12. Describe the construction of the skull. What is a suture? Tell how the peculiar form and structure of the skull adapt it for its use. Illustrate the impenetrability of the skull. Describe the experiment of the balls. What does it show?

13. What two cavities are in the trunk? Name its principal bones. Describe the spine. What is the object of the processes? Of the pads? Why is a man shorter at night than in the morning?

14. Describe the perfection of the spine. The articulation of the skull with the spine. Why is the atlas so called?

15. Describe the ribs. What is the natural form of the chest? Why is the thorax or chest made in separate pieces? How does the oblique position of the ribs aid in respiration? (See note, p. 80.)

16. How do the hip-bones give solidity? What two sets of limbs branch from the trunk? State their mutual resemblance.

17. Name the bones of the shoulder. Describe the collar-bone. The shoulder-blade. Can you describe the indirect articulation of the shoulder-blade with the trunk? Name the bones of the arm. Describe the shoulder-joint.

18. The elbow joint. The wrist.

19. Name the bones of the hand. The fingers. Describe their

---

* Bone structure may be summarized as follows: A bone is a collection of *Haversian elements* or rods. An Haversian element consists of a tube surrounded by *lamellæ*, which contain *lacunæ*, connected by *canaliculi.—Dr. T. B. Stowell.*

articulations. What gives the thumb its freedom of motion? In what lies the perfection of the hand?

20. Describe the hip-joint. What gives the upper limbs more freedom of motion than the lower? How does the pressure of the air aid us in walking? Illustrate. How do the gestures of the hand enforce our ideas and feelings?

21. Name the bones of the lower limbs. Describe the knee-joint. The patella. What is the use of the fibula? Can you show how the lower extremity of the fibula, below its juncture with the tibia, is prolonged to form a part of the ankle-joint? Name the bones of the foot. What is the use of the arch of the foot? What makes the step elastic?

22. Describe the action of the foot as we step. In graceful walking should the toes or the heel touch the ground first? What are the causes of deformed feet? What is the natural position of the big toe? Did you ever see a big toe lying in a straight line with the foot, as shown in statuary and paintings? How should we have our boots and shoes made? What are the effects of high heels? Of narrow heels? Of narrow toes? Of tight-laced boots? Of thin soles? What are the rickets? Cause of this disease? Cure?

23-24. Causes of spinal curvature? Cure? What is a felon? Cure? Cause of bow-legs? Cure? Is there any provision for remedying defects in the body? Name one. What is the correct position in sitting at one's desk? Is there any necessity for walking and sitting erect? Describe the bad effects of a stooping position. What is a sprain? Why does it need special care? What is a dislocation?

THE MUSCLES.

29. WHAT is the use of the skeleton? How is it concealed? Why is it the image of death? What are the muscles? How many are there? What peculiar property have they? Name other properties of muscles. *Ans.* Tonicity, elasticity.

30. How are they arranged? Where is the biceps? The triceps? How do the muscles move the limbs? Illustrate. What is the cause of squinting? Cure? (See p. 226.) Name and define the two kinds of muscles. Illustrate each. ·

31. What is the structure of a muscle? Of what is a fibril itself

composed? How does the peculiar construction of the muscle confer strength?

32. Describe the tendons. What is their use? Illustrate the advantages of this mode of attachment.

33. What two special arrangements of the tendons in the hand? Their use? How is the rotary motion of the eye obtained?

34. What is a lever? Describe the three classes of levers. Illustrate each. Describe the head as a lever. What parts of the body illustrate the three kinds of levers?

35. Give an illustration of the second class of levers. The third class. Why is the Tendon of Achilles so named? What is the advantage of the third class of levers? Why used in the hand? What class of lever is the lower jaw?

36. What advantages are gained by the enlargement of the bones at the joints? Illustrate. How do we stand erect? Is it an involuntary act? Why cannot a child walk at once, as many young animals do?

37. Why can we not hold up the head easily when we walk on "all fours"? Why cannot an animal stand erect as man does? Describe the process of walking. Show that walking is a process of falling.

38. Describe the process of running. What causes the swinging of the hand in walking? Why are we shorter when walking?* Why does a person when lost often go in a circle? In which direction does one always turn in that case?†

39. What is the muscular sense? Its value? Value of exercise?

40. Is there any danger of violent exercise? For what purpose should we exercise? Should exercise be in the open air? What is the rule for exercise? Is a young person excusable, who leads a sedentary life, and yet takes no daily out-door exercise? What will be nature's penalty for such a violation of her law? Will a postponement of the penalty show that we have escaped it? Ought a scholar to study during the time of recess? Will a promenade in the vitiated air of the school-room furnish suitable exercise? What is the best time for taking exercise?

* Stand a boy erect against a wall. Mark his height with a stick. Now have him step off a part of a pace, and then several whole paces. Next, let him close his eyes, and walk to the wall again. He will be perceptibly lower than the stick, until he straightens up once more from a walking position.

† Take several boys into a smooth grass lot. Set up a stick at a distance for them to walk toward. Test the boys, to find which are left-handed, or right-handed ; which left-legged or right-legged. Then blindfold the boys and let them walk, as they think, toward the mark. See who varies toward the right, and who turns to the left.

41. Who can exercise before breakfast? What are the advantages of the different kinds of exercise? Should we not walk more? What is the general influence upon the body of vigorous exercise?

42-3. State some of the wonders of the muscles. What is the St. Vitus's dance? Cure? What is the locked-jaw? Causes? The gout? Cause? Cure? The rheumatism? Its two forms?

44-5. Peculiarity of the acute? Danger? Is there any particular mode of treating it? What is the lumbago? Give instances. What is a ganglion? Its cure? A bursa?

THE SKIN.

49. WHAT are the uses of the skin? Describe its adaptation to its place. What is its function as an organ? Describe the structure of the skin. The sensitiveness of the cutis. The insensitiveness of the cuticle.

50. How is the skin constantly changing? The shape and number of the cells? Value of the cuticle? How is the cuticle formed? *Ans.* By secretion from the cutis.

51. What is the complexion? Its cause? Why is a scar white? What is the cause of "tanning"? What are freckles? Albinos? Describe the action of the sun on the skin. Why are the hairs and the nails spoken of under the head of the skin?

52. Uses of the hair? Its structure? What is the hair-bulb? What is it called? How does a hair grow? When can it be restored, if destroyed? What is the danger of hair-dyes? Are they of any real value?

53. How can the hair stand on end? How do horses move their skin? Is there any feeling in a hair? Illustrate the indestructibility of the hair.

54. What are the uses of the nails? How do the nails grow? What is the mucous membrane? Its composition?

55. The connective tissue? Why so called? What use does it subserve? What is its character?

56. How does the fat exist in the body? Its uses? State the various uses of membrane in the body. Where is there no fat? Why are the teeth spoken of in connection with the mucous membrane? Name and describe the four kinds of teeth.

57. What are the milk teeth? Describe them. What teeth appear first? When do the permanent teeth appear? Describe their growth. Which one comes first? Last?

58. Describe the structure of the teeth.

59. How are the teeth fitted in the jaw? Why do the teeth decay? What care should be taken of the teeth? What caution should be observed?

60. What are the oil glands? Use of this secretion? What are the perspiratory glands? State their number. Their total length.

61. What are the " pores" of the skin? What is the perspiration? What is the constitution of the perspiration? Illustrate its value. Name the three uses of the skin.

62. Illustrate the absorbing power of the skin. Why are cosmetics and hair-dyes injurious? What relation exists between the skin and the lungs? When is the best time for a bath? Why?

63. Value of friction? Should a bath be taken just before or after a meal? Is soap beneficial? What is the "reaction"? Explain its invigorating influence. How is it secured?

64. General effect of a cold bath? Of a warm bath? If we feel chilly and depressed after a bath, what is the teaching? Describe the Russian vapor bath. Why is the sea-bath so stimulating? How long should one remain in any bath?

65. How does clothing keep us warm? Explain the use of linen as an article of clothing. Cotton. Woolen. Flannel. How can we best protect ourselves against the changes of our climate?

66. What colored clothing is best adapted for all seasons? Value of the nap? Furs? Thick *vs.* thin clothing? Should we wear thick clothing during the day, and in the evening put on thin clothing? Can children endure exposure better than grown persons? What is the erysipelas? How relieved?

67. Dropsy? Corns? Cause? Cure? In growing nails? Cure? Warts? Cure?

6S. Chilblain? Cause? Preventive? Wens? Cure?

RESPIRATION AND THE VOICE.

73. NAME the organs of respiration and the voice. Describe the larynx. The epiglottis. The œsophagus. What is meant by food " going the wrong way " ?

74. Describe the vocal cords. Their use. How is sound pro-duced?

75. How are the higher tones of the voice produced? The lower? Upon what does loudness depend? A falsetto voice? What is the cause of the voice "changing"?

76. What is speech? Vocalization? Could a person talk without his tongue? Illustrate. How are talking-machines made?

77. How is *a* formed by the voice? What is *h*? Difference between a sigh and a groan? What vowel sounds are made in laughing? Does whistling depend on the voice? Tell how the various conso-nants are formed. What are the labials? The dentals? The lin-guals? What vowels does a child pronounce first?

78. Describe the wind-pipe. The bronchi. The bronchial tubes. Why is the trachea so called?

79. Describe the structure of the lungs. What are the lungs of slaughtered animals called? Why will a piece of the lungs float on water? Name the wrappings of the lungs. Describe the pleura. How is friction prevented? What are the cilia? What is their use?

80. What two acts constitute respiration? In what two ways may the position of the ribs change the capacity of the chest? Describe the process of respiration. Expiration. How often do we breathe? Describe the diaphragm.

81. What is sighing? Coughing? Sneezing? Snoring? Laugh-ing? Crying? Hiccough? Yawning? Its value?

82. What is meant by the breathing capacity? How does it vary? How much, in addition, can the lungs expel forcibly? How much of the breathing capacity is available only through practice? Value of this extra supply? Can we expel all the air from our lungs? Value of this constant supply?

83. How constant is the need of air? What is the vital element of the air? Describe the action of the oxygen in our lungs. What does the blood give up? Gain? How can this be tested? What are the constituents of the air? What are the peculiar properties and uses of each? What is the condition of the air we exhale? Which is the most dangerous constituent? Describe the evil effects of re-breathing the air.

84. For what is the "Black Hole of Calcutta" noted? Give other illustrations of the dangers of bad air. What is meant by the germs of disease floating in the air?

85-93. Describe the need of ventilation. Will a single breath pol-

lute the air? What is the influence of a fire or a light? Of a hot
stove? When is the ventilation perfect? What diseases are largely
owing to bad air? Should the windows and doors be tightly closed,
if we have no other means of ventilation? Is not a draught of air
dangerous? How can we prevent this, and yet secure fresh air?
What is the general principle of ventilation? Must pure air necessarily be cold air? Are school-rooms properly ventilated? What is
the effect? Are churches? Are our bed-rooms? Can we, at night,
breathe anything but night air? Is the night air out-of-doors ever
injurious? *Ans.* It is, in times and places of malaria, and also in
very damp weather, and should be avoided, even at the risk of bad air
in-doors.

93. Describe some of the wonders of respiration. How is constriction of the lungs produced?

94. When may clothing be considered tight? What are the dangers
of tight-lacing? Which would make the stronger, more vigorous, and
longer-lived person, the form shown in *A* or *B*, Fig. 33? Is it safe to
run any risk in this dangerous direction?

95. What is the bronchitis? Pleurisy? Pneumonia? Consumption? What is one great cause of this disease? How may a consti-
tutional tendency to this disease be warded off in youth? *Ans.* Besides plenty of fresh air and exercise, care should be taken in the diet.
Rich pastry, unripe fruit, salted meat, and acid drinks should be
avoided, and a certain quantity of fat should be eaten at each meal.—
Bennett. What is asphyxia? Describe the process for restoring such
a person. (See *Appendix.*) What is the diphtheria? Its peculiarity?
Danger?

96. The croup? Its characteristics? Remedy? (See *Appendix.*)
Causes of stammering? How cured?

THE CIRCULATION.

103. NAME the organs of the circulation. Does the blood permeate all parts of the body? What is the average amount in each
person? Its composition? The plasma? The red corpuscles? The
white?

104. What is the size of a red cell? Are the shape and size uniform? Value of this? Illustrate. Are the disks permanent?

105. What substances are contained in the plasma? What is fibrin? In what sense is the blood "liquid flesh"? What is the use of the red disks? What is the office of the oxygen in the body? Where is the blood purified?

106. What is transfusion? Give some illustrations. Is it of value?

107. What is the cause of coagulation of the blood? Value of this property? Has the fibrin any other use? What organ propels the blood?

108. What is the location of the heart? How large is it? Put your hand over it. What is the pericardium?

109. Describe the systole. The diastole. How many chambers in the heart? What is their average size? What is meant by the right and the left heart?

110. What are the auricles? Why so called? The ventricles? What is the use of the auricles? The ventricles? Which are made the stronger? Show the need of valves in the ventricles. Why are there no valves in the auricles? Draw on the board the form of the valves. Name them.

111. Describe the tricuspid valve. The bicuspid.

112. How are these valves strengthened? What peculiarity in the attachment of these cords? Describe the semi-lunar valves. What are the arteries? Why so named? What is their use? Their structure? How does their elasticity act? What is meant by a "collateral circulation"?

113. How are the arteries protected? Where are they located? Give a general description of the arterial system. What is the aorta? What is the pulse? On which arteries can we best feel it? What is the average number of beats per minute? How and why does this vary?

114. Why does a physician feel a patient's pulse? What are the veins? What blood do they carry? Describe the venous system. What vein does not lead toward the heart?

115. Describe the valves of the veins. What valves of the heart do they resemble? Where and how can we see the operation of these valves? What are the capillaries? What is the function of the capillaries?* What changes take place in this system? What are varicose veins?

---

* The distinctive function of the capillaries is to offer peripheral resistance to the circulation of the blood. This insures "blood pressure," a condition indispensable to the "heart beat," and also causes leakage (transudation). This leakage brings the nutriment in contact with the tissue cells, whereby they are renewed. In the same way the air passes from the blood to the cells.

116. Describe the circulation of the blood as seen in the web of a frog's foot. In what two portions is the general circulation divided? Who discovered the circulation of the blood? How was the discovery received? What remark did Harvey make? What does that show?

117. Describe the route of the blood by the diagram? 1. The lesser circulation; 2. The greater circulation. What is the velocity of the blood? How long does it require for all the blood to pass through the heart? How has this been estimated? What is the shortest route the blood can take? The longest? How long does it take the blood to make the tour of the body? How has this been estimated? What is the average temperature of the body? How much does this vary in health? *Ans.* Not more than 2°, even in the greatest extremes of temperature.—*Flint.*

119. How and where is the heat of the body generated? How is it distributed? In what diseases is the variation of temperature marked? How is the temperature of the body regulated?

120. In what way does life exist through death? Is not this as true in the moral as in the physical world? What does it teach? How rapidly do our bodies change? What are the three vital organs?

121. Name some of the wonders of the heart.

122-4. What is the lymphatic circulation? What is the thoracic duct? The lymph? The glands? What is the office of the lymphatics? What are the lacteals? Give some illustrations of the action of the lymphatics of the different organs. Should we use care in selecting wall-paper? What is meant by the sub-cutaneous insertion of morphine? How do hibernating animals live during the winter? What is a congestion? Its cause? Blushing? Why does terror cause one to grow cold and pale?

125. How is an inflammation caused? Name its four characteristics. How may severe bleeding be stopped? How can you tell whether the blood comes from an artery or a vein? Why should you know this?

126-7. What is the scrofula? What are "kernels"? How may a scrofulous tendency of the system be counteracted? What kinds of food stimulate this disease? What is the cause of "a cold"? Why does exposure sometimes cause a cold in the head, sometimes on the lungs, and at others brings on a rheumatic attack? Why is a cold dangerous? *Ans.* It weakens the system and paves the way for other diseases. What is the theory of treating a cold? Describe the method. What is catarrh? Cause?

128. Illustrate the general effect of alcohol upon the circulation. Upon the heart. Upon the membrane. Upon the blood. Upon the lungs. What is the active principle of all liquors?

DIGESTION AND FOOD.

139. WHY do we need food? Why will a person starve without food? Are the current stories of people who live without food to be relied upon? How much food is needed per day by an adult in active exercise? How much in a year? How does this amount vary?

140. Describe the body as a mold. As an eddy. What does food do for us? What does food contain? How is this force set free? What force is this? How can it be turned into muscular motion, mental vigor, etc.? Do we then draw all our power from nature? What becomes of these forces when we are done with them? Do we destroy the force we use? *Ans.* No matter has been destroyed, so far as we know, since the creation, and force is equally indestructible.

141. Compare our food to a tense spring. What three kinds of food do we need? What is nitrogenous food? Name the common forms. What is the characteristic of nitrogenous food? Why called albuminous?

142. What is carbonaceous food? Its two kinds? Constituents of sugar? Where are starch and gum ranked? Why? Use of carbonaceous food? What becomes of this heat? Composition of fat? How does fat compare with sugar in producing heat? Name the other uses of carbonaceous food. From what kind of food does the body derive the greatest strength?

143. Name the mineral matters which should be contained in our food. What do you say of the abundance and necessity of water? Ought we not to exercise great care in selecting the water we drink?*

---

* Water which has passed through lead-pipes is apt to contain salts of that metal, and is therefore open to suspicion. Metallic-lined ice-pitchers, galvanized-iron reservoirs, and many soda-water fountains, are liable to the same objection. There are also organic impurities in water equally dangerous. River-water often disseminates the germs of typhoid fever and other diseases just as the air scatters the seeds of small-pox and scarlet fever. Thus the great outbreak of cholera in the east of London, in 1866, was traced to the contamination of the River Lea, which furnished the supply of water to that part of the city. The surface water frequently flows into a well carrying organic matter to poison its contents. Wells sometimes receive underground the drainage from grave-yards, manufactories, cess-pools, swamps, barn-yards, vaults, etc., all of which render the water unfit for use.

Will not the character of our food influence the quantity of water we need? What are the uses of these different minerals? Illustrate the importance of salt.

144. Could a person live on one kind of food alone? Illustrate. Describe the effect of living on lean meat. Show the necessity of a mixed diet. Illustrate. Show the need of digestion. Illustrate.

145. What is assimilation? Describe the general plan of digestion. What did Berzelius call digestion? Why? What amount of liquid is daily secreted by the alimentary canal?

146. What is the alimentary canal? How is it lined? How does the amæba digest its food? The hydra? Define secretion. Describe the saliva. How is it secreted? What is the amount?

147. Its organic principle? Its use? How soon does it act? How long? What tends to check or increase the flow of saliva?

149. Describe the process of swallowing. The stomach. Its size. Its construction. What is the peristaltic movement? The pylorus? For what does this open? What is the gastric juice? How abundant is it? To what is its acidity due?

150. What organic principle does it contain? How is its flow influenced? What is its use? Appearance of the food as it passes through the pylorus? How is pepsin prepared? Why is not the stomach itself digested? What is the construction of the intestines?

151. How are the intestines divided? What is the duodenum? Why so called? What juices are secreted here? What is the bile? Describe the liver. What is its weight? Its construction? *Ans.* It consists of a mass of polyhedral cells only $\frac{1}{100}$ to $\frac{1}{3000}$ of an inch in diameter, filling a mesh of capillaries. The capillaries carry the blood to and fro, and the cells secrete the bile. What is the cyst? What does the liver secrete from the blood besides the bile? Is the bile necessary to life? Illustrate. What is its use? What is the pancreatic juice? Its organic principle? Its use?

152. Appearance of the food when it leaves the duodenum? Describe the small intestine.

153. What is absorption? In what two ways is the food absorbed? Where does the process commence? How long does it last? Describe the lacteals. Of what general system do they form a part? What do the veins absorb? Where do they carry the food?

154. How is it modified? Describe the complexity of the process of digestion. What length of time required for digestion in the stomach?

155. May not food which requires little time in the stomach need more in the other organs, and *vice versa*? Tell the story of Alexis St. Martin. What time was required to digest an ordinary meal? Apples? Eggs, raw and cooked? Roast beef? Pork? Which is the king of the meats? What is the nutritive value of mutton? Lamb? How should it be cooked? Objection to pork? What is the trichina? Should ham ever be eaten raw?

156. Value of fish? Oysters? Milk? Cheese? Eggs? Bread? Brown bread? Are warm biscuit and bread healthful? Nutritive value of corn? The potato? Of ripe fruits?

157. Of coffee? To what is its stimulating property due? Its influence on the system? When should it be discarded? Should children use any stimulants? Effects of tea? Influence of strong tea? What is the active principle of tea?

158. Nutritive value of chocolate? What is its active principle? Story of Linnæus? How should tea be made? What is the effect of cooking food? What precaution in boiling meat? In roasting? Object of this high temperature? What precaution in making soup? Why is frying an unhealthful mode of cooking?

159. State the five evil results of rapid eating. What disease grows out of it? If one is compelled to eat a meal rapidly, as at a railroad station, what should he take? Why? Why does a child need more food proportionately than an old person? State the relation of waste to repair in youth, in middle, and in old age. What kind and quantity of food does a sedentary occupation require? What caution should students who have been accustomed to manual labor observe? Must a student starve himself? Is there not danger of over-eating? Would not an occasional abstinence from a meal be beneficial? Do not most people eat more than is for their good? How should the season regulate our diet?

160. The climate? Illustrate. What does a natural appetite indicate? How are we to judge between a natural and an artificial longing? What does the craving of childhood for sugar indicate?* What

---

* It does not follow from this, however, that the free use of sugar in its separate form is desirable. The ordinary articles of vegetable food contain sugar (or starch, which in the body is converted into sugar), in large proportion; and there is good reason to believe that in its naturally-combined form it is both more easily digested, and more available for the purposes of nutrition, than when crystallized. The ordinary sugar of commerce, moreover, derived from the sugar-cane, is not capable of being directly applied to physiological purposes. Cane-sugar is converted within the body into another kind of sugar, identical with that derived from the grape, before it can enter into the circuit of the vital changes.

is the effect upon the circulation of taking food? Should we labor or study just before or after a meal? Why not? What time should intervene between our meals? Is "lunching" a healthful practice? Eating just before retiring?

161. Why should care be banished from the table? Will a regular routine of food be beneficial? Describe some of the wonders of digestion.

162-3. What are the principal causes of dyspepsia? How may we avoid that disease? What are the mumps? What care should be taken? Is alcohol a food? Illustrate. Compare the action of alcohol with that of water. Is all the alcohol taken into the stomach eliminated unchanged? Does alcohol contain any element needed by the body? What is the effect of alcohol upon the digestion? Will pepsin act in the presence of alcohol? What is the effect of alcohol upon the liver? What is "Fatty Degeneration"? What is the effect of alcohol upon the kidneys? Does alcohol impart heat to the body? Does it confer strength? What does Dr. Kane say? Describe Richardson's experiments. Tell what peculiar influence alcohol exerts. What is alcoholism? What is heredity?

THE NERVOUS SYSTEM.

177. WHAT are the organs of the nervous system? What is the general use of this system? How does it distinguish animals from plants? What are the vegetative functions? What is the gray matter? Its use? The white matter? Its use?

179. Describe the brain. What is its office? Its size? How does it vary? Illustrate. Name its two divisions. Describe the cerebrum.

180. The convolutions. The membranes which bind the brain together. What do you say of the quantity of blood which goes to the brain? What does it show? What do the convolutions indicate?

181. What is the use of the two halves of the brain? What theories have been advanced concerning it? What is the effect of removing the cerebrum?

182. Describe the cerebellum. What is the arbor vitæ? What does this part of the brain control? What is the effect of its being injured? Illustrate.

183. Describe the spinal cord. What is the medulla oblongata? Describe the nerves. Is each part of the body supplied with its own nerve? Prove it.

184-5. Name the three classes of nerves. What are the motory nerves? The sensory? When will motion be lost and feeling remain, and *vice versa?* What is meant by a transfer of pain? Illustrate. What are the spinal nerves? Describe the origin of the spinal nerve. What are the cranial nerves?

185-187. Describe the sympathetic system. What is its use? How does the brain control all the vital processes? What is meant by the crossing of the cords?

188. What is the effect? What exception in the seventh pair of cranial nerves? What is reflex action? Give illustrations.

189. Give instances of the unconscious action of the brain.* Can

---

* Dr. Carpenter, in the course of a recent lecture at Manchester, England, upon the " Unconscious Action of the Brain," gave the following among other illustrations:

1. We find that when we set off in the morning with the intention of going to our place of employment, not only do our legs move without our consciousness, if we are attending to something entirely different, but we guide ourselves in our walk through the streets; we do not run up against anybody we meet; we do not strike ourselves against the lamp-posts; and we take the appropriate turns which are habitual to us. It has often happened to myself, and I dare say it has happened to every one of you, that you have intended to go somewhere else—that when you started you intended instead of going in the direct line to which you were daily accustomed, to go a little out of your way to perform some little commission; but you have got into a train of thought and forgotten yourself, and you find that you are half-way along your accustomed track before you become aware of it. Now, there, you see, is the same automatic action of these sensory ganglia—we see. we hear—for instance, we hear the rumbling of the carriages, and we avoid them without thinking of it—our muscles act in respondence to these sights and sounds—and yet all this is done without our intentional direction— they do it for us. We arrive at a certain point where we are accustomed to stop. and are surprised that we have reached it. You will ask me, perhaps, " What is the exciting cause of this succession of actions in walking?" I believe it is the contact of the ground with the foot at each movement. We put down the foot, that suggests as it were to the spinal cord the next movement of the leg in advance, and that foot comes down in its turn ; and so we follow with this regular rhythmical succession of movements. It is all done through the reflex action of the spinal cord.

2. The cerebellum has its unconscious action in the processes of respiration, the involuntary movements which are made in response to the senses, as in winking. starting back at a sound, etc.

3. The cerebrum acts automatically in cases familiar to all. A large part of our mental activity consists of this unconscious work of the brain. There are many cases in which the mind has obviously worked more clearly and more successfully in this automatic condition, when left entirely to itself. than when we have been cudgeling our brains. so to speak, to get the solution. An instance was put on record by a gentleman well known in London, the Rev. John De Liefde, a Dutch clergyman, who gave it on the authority of a fellow-student who had been at the college

there be feeling or motion in the lower limbs when the spinal cord is destroyed? What does the story told by Dr. John Hunter show? Give illustrations of the independent action of the spinal cord in animals. What are the uses of reflex action?

190. State its value in the formation of habits. How does the brain grow? What laws govern it? What must be the effect of constant light-reading? Of over-study or mental labor?

191. State the relation of sleep to repair and waste. How many hours does each person need? What kind of work requires most sleep?

192. What is the influence of sunlight on the body? Illustrate. Name some of the wonders of the brain.

193–8. What four stages are there in the effect of alcohol on the nervous system? Describe each. Does alcohol confer any permanent strength? What is the physiological effect of alcohol on the brain? On the mental and moral powers? What is the Delirium Tremens?

198. What are the principal constituents of tobacco? Should a man be punished for a crime he commits while drunk?

199. What are the physiological effects of tobacco?

200. Who are most likely to escape injury?

201. Is tobacco a food? What is its influence upon youth? Why are cigarettes specially injurious? What effect does tobacco have on the sensibilities?

202. Name illustrations of the injurious effect of tobacco on young men. How is opium obtained? What is its physiological effect? Can one give up the use of opium when he pleases? What is the harmful influence of chloral hydrate? Of chloroform?

at which he studied in early life. He had been attending a class in mathematics, and the professor said to his students one day: "A question of great difficulty has been referred to me by a banker—a very complicated question of accounts, which they have not themselves been able to bring to a satisfactory issue, and they have asked my assistance. I have been trying, and I cannot resolve it. I have covered whole sheets of paper with calculations, and have not been able to make it out. Will you try?" He gave it to them as a sort of problem, and said he should be extremely obliged to any one who would bring him the solution by a certain day. This gentleman tried it over and over again; he covered many slates with figures, but could not succeed in resolving it. He was " put on his mettle," and determined to achieve the result. But he went to bed on the night before the solution was to be given in, without having succeeded. In the morning, when he went to his desk, he found the whole problem worked out in his own hand. He was perfectly satisfied that it was his own hand; and this was a very curious part of it – that the result was correctly obtained by a process very much shorter than any he had tried. He had covered three or four sheets of paper in his attempts, and this was all worked out upon one page, and correctly worked, as the result proved.

THE SPECIAL SENSES.

211. WHAT is a sense? Name the five senses. To what organ do all the senses minister? If the nerve leading to any organ of sense be cut, what would be the effect?* Sometimes persons lose feeling in a limb, but retain motion : why is this? What is the sense of touch sometimes called? Describe the organ of touch. What are the papillæ? Where are they most abundant?† What are the uses of this sense? What special knowledge do we obtain by it? Why do we always desire to handle anything curious?

212. Can the sense of touch always be relied upon? Illustrate. What is the *tactus eruditus?* Illustrate. Tell how one sense can take the place of another. Give illustrations of the delicacy of touch possessed by the blind.

213. Describe the sense of taste. How can you see the papillæ of taste?

214. What causes the velvety look of the tongue? Why do salt and bitter flavors induce vomiting? Why does an acid "pucker" the face? What substances are tasteless? Illustrate. Has sulphur any taste? Chalk? Sand? What is the use of this sense? Does it not also add to the pleasures of life? Why are the acts of eating, drinking, etc., thus made sources of happiness? Describe the organ of smell. State the intimate relation which exists between the senses of smell and taste. Name some common mistakes which occur in consequence.

215. Must the object to be smelled touch the nose? What is the theory of smell? How do you account for the statement made in the note concerning musk and ambergris?

216. What are the uses of this sense? Are agreeable odors healthful, and disagreeable ones unhealthful? Describe the organ of hearing. Describe the external ear. What is the tympanum or drum of the ear?

* Each organ is adapted to receive a peculiar kind of impression. Hence we cannot smell with the eyes nor see with the nose. So that if the nerve communicating between the brain and any organ be destroyed, that means of knowledge is cut off.

† If we apply the points of a compass blunted with cork to different parts of the body, we can distinguish the two points at one-twenty-fourth of an inch apart on the tongue, one-sixteenth of an inch on the lips, one-twelfth of an inch on the tips of the fingers, and one-half inch on the great toe ; while, if they are one inch on the cheek, and two inches on the back, they will scarcely produce a separate sensation. —*Huxley.*

217. Describe the middle ear. Name the bones of the ear. Describe their structure. Describe the internal ear. By what other name is it known? What substances float in the liquid which fills the labyrinth? What is their use? Describe the fibers of Corti. What do they form? Use of this microscopic harp?

218. Give the theory of sound. Where is the sound, in the external object or in the mind? Can there be any sound, then, where there is no mind? What advice is given concerning the care of the ear? How can insects be removed? Which sense would you rather lose, hearing or sight? Does not a blind person always excite more sympathy than a deaf one? How does the sight assist the hearing?*

219. Describe the eye. Name the three coats of which it is composed. Is it a perfect sphere? *Ans.* The cornea projects in front, and the optic nerve at the back sticks out like a handle, while the ball itself has its longest diameter from side to side.

220. How is the interior divided? Object of the crystalline lens? How is the crystalline lens kept in place? Describe the liquids which fill the eye. What is the pupil? Describe the eyelids. Why is the inner side of the eyelid so sensitive? What is the cause of a black eye? Use of the eyelashes?

221. Where are the oil glands located? What is their use? Describe the lachrymal gland. The lachrymal lake. What causes the overflow in old age? Explain the structure of the retina. Use of the rods and cones. What is the blind spot? Illustrate.

222-6. What is the theory of sight? Illustrate. State the action of the crystalline lens. Its power of adaptation. Cause of nearsightedness. How remedied? Cause of far-sightedness? How remedied? Do children ever need spectacles? What is the cataract? How cured? What care should be taken of the eyes? Should one constantly lean forward over his book or work? What

---

* In *hearing*, the attention is more or less characteristic. If we wish to distinguish a distant noise, or perceive a sound, the head inclines and turns in such a manner as to present the external ear in the direction of the sound, at the same time the eyes are fixed and partially closed. The movement of the lips of his interlocutor is the usual means by which the deaf man supplies the want of hearing ; the eyes and the entire head, from its position, having a peculiar and painful expression of attention. In looking at the portrait of La Condamine, it was easily recognized as that of a deaf person. Even when hearing is perfect, the eyes act sometimes as auxiliaries to it. In order to understand an orator perfectly, it seems necessary to see him—the gestures and the expression of the face seeming to add to the clearness of the words. The lesson of a teacher cannot be well understood if any obstacle is interposed between him and the eyes of the listening pupil. So that if a pupil's eyes wander, we know that he is not attentive.—*Wonders of the Human Body.*

special care should near-sighted children take? By what carelessness may we impair our sight? Should we ever read or write at twilight? Danger of reading upon the cars? What course should we take when objects get into the eye? How may they be removed? Are "eye-stones" useful? Why we should never use eye-washes except upon the advice of a competent physician? What care should be taken with regard to the direction of the light when we are at work?

CONCLUSION.

STATE some of the benefits of health. Contrast it with sickness. How were diseases formerly supposed to be caused? What remedies were used? What does modern science teach us to be the nature of disease? Give some illustrations showing how diseases may be prevented. Is it probable that the body was intended to give out in any one of its organs? What is the first step to be taken in the cure of a disease? What should be the object of medicine? What is now the chief dependence of the best physicians? What do you think concerning the common use of patent nostrums? Relate the story told of the remedy employed by the African medicine-man. Ought we not to use the greatest care in the selection of our physician, to secure the highest medical skill and cultivation?

INDEX.